Science, Technology and Medicine in Modern History

General Editor:

John V. Pickstone, Centre for the History of Science, Technology and Medicine, University of Manchester, England (www.man.ac.uk/CHSTM)

One purpose of historical writing is to illuminate the present. At the start of the third millennium, science, technology, and medicine are enormously important, yet their development is little studied.

The reasons for this failure are as obvious as they are regrettable. Education in many countries, not least in Britain, draws deep divisions between the sciences and the humanities. Men and women who have been trained in science have too often been trained away from history, or from any sustained reflection on how societies work. Those educated in historical or social studies have usually learned so little of science that they remain thereafter suspicious, overawed, or both.

Such a diagnosis is by no means novel, nor is it particularly original to suggest that good historical studies of science may be peculiarly important for understanding our present. Indeed, this series could be seen as extending research undertaken over the last half-century. But much of that work has treated science, technology, and medicine separately; this series aims to draw them together, partly because the three activities have become ever more intertwined. This breadth of focus and the stress on the relationships of knowledge and practice are particularly appropriate in a series that will concentrate on modern history and on industrial societies. Furthermore, while much of the existing historical scholarship is on American topics, this series aims to be international, encouraging studies on European material. The intention is to present science, technology, and medicine as aspects of modern culture, analysing their economic, social, and political aspects, but not neglecting the expert content that tends to distance them from other aspects of history. The books will investigate the uses and consequences of technical knowledge, and how it was shaped within particular economic, social and political structures.

Such analyses should contribute to discussions of present dilemmas and to assessments of policy. 'Science' no longer appears to us as a triumphant agent of Enlightenment, breaking the shackles of tradition, enabling command over nature. But neither is it to be seen as merely oppressive and dangerous. Judgement requires information and careful analysis, just as intelligent policy-making requires a community of discourse between men and women trained in technical specialities and those who are not.

This series is intended to supply analysis and to stimulate debate. Opinions will vary between authors; we claim only that the books are based on searching historical study of topics which are important, not least because they cut across conventional academic boundaries. They should appeal not just to historians, nor just to scientists, engineers and doctors, but to all who share the view that science, technology and medicine are far too important to be left out of history.

Thomas Schlich
SURGERY, SCIENCE AND INDUSTRY
A Revolution in Fracture Care, 1950s–1990s

Eve Seguin (*editor*)
INFECTIOUS PROCESSES
Knowledge, Discourse and the Politics of Prions

Crosbie Smith and Jon Agar (*editors*)
MAKING SPACE FOR SCIENCE
Territorial Themes in the Shaping of Knowledge

Stephanie J. Snow
OPERATIONS WITHOUT PAIN
The Practice and Science of Anaesthesia in Victorian Britain

Carsten Timmermann
A HISTORY OF LUNG CANCER
The Recalcitrant Disease

Carsten Timmermann and Julie Anderson (*editors*)
DEVICES AND DESIGNS
Medical Technologies in Historical Perspective

Carsten Timmermann and Elizabeth Toon (*editors*)
CANCER PATIENTS, CANCER PATHWAYS
Historical and Sociological Perspectives

Jonathan Toms
MENTAL HYGIENE AND PSYCHIATRY IN MODERN BRITAIN

Marius Turda
EUGENICS AND NATION IN EARLY 20TH CENTURY HUNGARY

Duncan Wilson
TISSUE CULTURE IN SCIENCE AND SOCIETY
The Public Life of a Biological Technique in Twentieth Century Britain

---

Science, Technology and Medicine in Modern History
Series Standing Order ISBN 978–0–333–71492–8 hardcover
Series Standing Order ISBN 978–0–333–80340–0 paperback
(*outside North America only*)

You can receive future titles in this series as they are published by placing a standing order. Please contact your bookseller or, in case of difficulty, write to us at the address below with your name and address, the title of the series and one of the ISBNs quoted above.

Customer Services Department, Macmillan Distribution Ltd, Houndmills, Basingstoke, Hampshire RG21 6XS, England

---

The Changing Faces of Childhood Cancer

Clinical and Cultural Visions since 1940

Emm Barnes Johnstone
Teaching and Research Fellow, Royal Holloway, University of London, UK

with

Joanna Baines
Freelance Academic Copyeditor, UK

First published 2015 by
PALGRAVE MACMILLAN

Palgrave Macmillan in the UK is an imprint of Macmillan Publishers Limited, registered in England, company number 785998, of Houndsmills, Basingstoke, Hampshire, RG21 6XS.

Palgrave Macmillan in the US is a division of St Martin's Press LLC, 175 Fifth Avenue, New York, NY 10010.

Palgrave is the global academic imprint of the above companies and has companies and representatives throughout the world.

Palgrave® and Macmillan® are registered trademarks in the United States, the United Kingdom, Europe and other countries.

ISBN 978-1-349-54185-0   ISBN 978-1-137-35352-8 (eBook)

DOI 10.1057/9781137353528

A catalogue record for this book is available from the British Library.

A catalog record for this book is available from the Library of Congress.

Typeset by Mps Limited, Chennai, India.

Transferred to Digital Printing in 2014

*We dedicate this book to John Pickstone,
died 2014, colleague, mentor, friend*

Contents

List of Figures

Acknowledgements

This book would not have been possible without the support of the many pioneering paediatric oncologists, pathologists, epidemiologists, and fundraisers who allowed me to interview them and made available their collections of papers. I want to thank, in particular, Judith Chessels, Tim Eden, Jim Malpas, Henry Basil Marsden, and Pat Morris Jones. Pamela Barnes shared her personal archives on the development of specialist wards and playrooms for child patients. Sue Ablett gave me access to the archives of the Children's Cancer and Leukaemia Group, formerly the UK Children's Cancer Study Group. Gordon Piller, Mark Wilson, and Pete Crowley at Leukaemia and Lymphoma Research, formerly the Leukaemia Research Fund, kindly granted me permission to look through archives from the charity's early history. I would also like to thank Janine Allis-Smith of Cumbrians Opposed to a Radioactive Environment, Amanda Engineer at the Wellcome Library and archivists at the John P. McGovern Historical Collections and Research Center of the Houston Academic of Medicine Texas Medical Center Library, as well as the archivists at the National Archives, Kew.

I am grateful to former colleagues Helen Valier and Elizabeth Toon for comments on early versions of many of these chapters, and Ilana Löwy and Carsten Timmermann for detailed comments on the entire manuscript. Finally, I am especially grateful to Jo Baines, who helped take an idea and turn it into reality.

Emm Barnes Johnstone

I would like to thank my former colleagues from the Centre for the History of Science, Technology and Medicine at the University of Manchester, staff at The Christie, particularly Anne Murtagh, and, of course, Emm Barnes Johnstone.

Joanna Baines

About the Authors

Emm Barnes Johnstone is a Teaching and Research Fellow in the History Department at Royal Holloway, University of London. Her research in the history of medicine has covered nineteenth-century folk psychology, twentieth century developments in care for the chronically ill and for children, and the history of rare cancers since the 1930s. She has worked extensively with museums and universities to establish programmes of events and exhibitions that engage diverse audiences with material history and with the history of ideas. Recent publications include (with Julie Anderson and Emma Shackleton) *The Art of Medicine: Over 2000 Years of Images and Imagination* (2011).

Joanna Baines completed her doctorate on the individualisation of cancer at the Centre for the History of Science, Technology and Medicine, University of Manchester, in 2010. She has devised and produced historical public engagement materials and events, and provides editing services for academics.

List of Abbreviations

ABCC	Atomic Bomb Casualty Commission
BECC	British Empire Cancer Campaign
BNFL	British Nuclear Fuels Limited
BPA	British Paediatric Association
CAM	complementary and alternative medicine
CNS	central nervous system
CRC	Cancer Research Campaign
CSTG	Children's Solid Tumour Group
DSM-IV	Diagnostic and Statistical Manual of Mental Disorders, Fourth Edition
GP(s)	general practitioner(s)
ICRF	Imperial Cancer Research Fund
LRF	Leukaemia Research Fund
MP	Member of Parliament
MRC	Medical Research Council
NCCS	National Coalition for Cancer Survivorship
NCI	National Cancer Institute
NHL	non-Hodgkin's lymphoma
NHS	National Health Service
NIH	National Institutes of Health
NRC	National Research Council
NRPB	National Radiological Protection Board
RCN	Royal College of Nursing
RCTs	Randomised Controlled Trials
SIOP	Société Internationale d'Oncologie Pédiatrique
SKI	Sloan Kettering Institute
SWCCSG	Southwest Cancer Chemotherapy Study Group
UICC	Union Internationale Contre le Cancer
UKCCSG	United Kingdom Children's Cancer Study Group
YTV	Yorkshire Television

Introduction

Cancer in children is rare. Cases of cancer in people younger than 15 years of age (the commonly accepted definition of childhood amongst cancer specialists in Great Britain) accounts for only half of one per cent of all cancers recorded in Britain, with around 1,600 new patients diagnosed each year. While two-thirds of these are cases of just a few types – leukaemia, brain and central nervous system tumours, and lymphomas – cancer takes many other forms in children, including tumours in the kidney, eye, bone, liver, skin, soft tissue, and reproductive organs. Some of these conditions are so uncommon that only a handful of new cases are discovered in Great Britain each year (Cancer Research, 2012).

Despite this scarcity, the childhood cancer patient is a familiar cultural icon in twenty-first century Great Britain, a symbol of innocent victimhood that has inspired many marathons to be run and healthy heads to be shaved to raise money. This has not always been the case. Unlike in the USA where, as Gretchen Krueger has described, 'poster children' were used to both convey information about treatment and evoke emotional and financial support from the 1930s onwards, this phenomenon did not appear in Great Britain until the late 1960s (Krueger, 2008). This delay was only partly due to the fabled British reserve: a reticence to discuss cancer that thwarted native attempts to replicate the American cancer prevention initiatives of the early twentieth century (Patterson, 1991, p. 142; Cantor, 2007, pp. 9–10; Toon, 2007, pp. 120–2). The relative scarcity of childhood cancers, compounded by the fact that many went undiagnosed, meant both the public and the medical profession were largely unfamiliar with these conditions, and no specialist charities promoted awareness of these diseases. In addition, although historians have demonstrated that there was a reconceptualisation of malignant

1

disease in general as a potentially curable condition during the late nineteenth century, a similar transformation concerning childhood cancer was a more complex and prolonged process, one that continued into the late twentieth century (Lerner, 2001; Pinell, 2002). Until the 1950s, acute leukaemia was invariably classed as a fatal illness, and the majority of childhood solid tumours that had undergone metastasis (i.e. had spread to other parts of the body) were seen, at the very least, as unmanageable. For those who had come across these conditions in the first half of the century, causes for optimism were scarce.

This book originates from the Centre for the History of Science, Technology and Medicine at the University of Manchester. The Centre was awarded a Wellcome Trust Programme Grant to study the history of cancer research and services in Great Britain since 1945, and this book will draw on work resulting from the initiative by our former colleagues, John Pickstone, Elizabeth Toon, Carsten Timmermann, and Helen Valier. We examine the development of childhood cancer treatment and research in Great Britain since the 1930s – a time when only two medical practitioners specialised in the subject – through the processes of problem recognition, data collection, the development of cooperative groups and clinical trials, and the centralisation of services and professionalisation of the field. We will discuss these developments in relation to those in other countries, the USA in particular, where some innovations in research and treatment occurred earlier, watched and welcomed by British practitioners. However, the development of childhood cancer services in Great Britain was not delayed; rather, it followed a different path from that seen in the USA. Owing to differences in population size, in the financial and structural basis of the medical systems, in specialisation and discipline interaction, and in general cultural attitudes, the British story is unique.

The dramatic transformation of childhood cancer from an almost inevitably fatal rarity in the 1940s to an often survivable experience by the 1990s has attracted many scholars of biomedical research, interested in studying the work of clinicians and laboratory-based scientists who worked together, across hospitals and across countries, to develop not only new treatments, but also novel ways of conducting biomedical research. Keating and Cambrosio have presented a detailed account of the emergence of cooperative groups for the administration of clinical trials into cancer treatments in the USA and in Europe. Quirke, Gaudillière, and Löwy, and our former colleagues from Manchester, have also provided rich analyses of the dynamics driving new behaviours in biomedical research – and the impact of these on methods for communicating information about treatment success rates, as well as changes in patient outcomes – comparing national contexts and variations. The work of these historians is drawn on in the chapters that follow, as we look specifically at changing approaches to childhood leukaemias, lymphomas, and tumours (Gaudillière, 2006; Keating and Cambrosio, 2002, 2007, 2012; Löwy, 1996; Pickstone, 2007; Quirke and Gaudillière, 2008; Valier and Timmermann, 2008).

In this book we wish to build on and from institutional history, looking at the relation between changing institutions and the meanings of a disease. We examine how the overlapping histories of innovation and adaptation in biomedical research institutions translated into differing and evolving cultural meanings of childhood cancer in Great Britain. We consider how events as far away from home as Uganda could inspire the media and excite the public imagination; we explore the tension between a medical profession reluctant to entertain dubious claims and a public keen to explore alternative solutions to the cancer problem; and, most importantly, we address the patients themselves, both those who lived at a time when optimism was largely unfounded, and the growing number who have been cured – patients whose survival has added a field of study for cancer researchers.

The individual experiences of childhood cancer that make up our subject matter include the feelings and reminiscences of professionals who have sought to measure, understand, and eradicate this family of diseases, as recorded in their publications and in interviews; accounts of childhood cancer from patients and parents who have lived with the messy realities of treatment; and stories collected by journalists covering individual patient pathways or medical 'breakthroughs'. This focus on collections of experiences brings together our interests in anthropology and cultural history, foregrounding narratives about the meaning of cancer in the young, of clinical experimentation, of remission, and the problematic nature of survival. Our account is also shaped by personal experiences; one of us has had cancer, the other is mother to a survivor of childhood cancer.

The voices of children are often hard for historians to locate and recapture as their viewpoints are rarely stored in archives or seen as potentially relevant for future generations. However, childhood cancer proves an exception: as the expected outcomes for sufferers proved capable of altering rapidly as treatments were modified, patients' experiences *were* recorded, by medical professionals, families, and the media. Thus, we have a surprisingly rich bank of individual pathways through cancer available for consideration. As Chris Feudtner wrote in his history of diabetes, the history of a disease 'is not one story but many: all the separate stories of patients' (2003, p. xiv). And so this book traces two intimately related developments: firstly, the emergence of childhood cancer as a cluster of visible, interesting, and concerning diseases, best investigated and treated by a unified community; and, secondly, the shifting and settling meanings of childhood cancer expressed in patient stories.

In Chapter 1 we examine the knowledge, or lack thereof, of childhood cancers in Great Britain during the 1930s. In the USA, haematologists (specialists in blood diseases) and pharmaceutical companies would dominate the early development of paediatric oncology. In Great Britain, however, where the field of radiotherapy was comparatively advanced and that of paediatrics relatively unorganised, it was two individuals initially trained as paediatricians with a keen interest in radiotherapy who would lead the field: Ivor G. Williams and Edith Paterson. In the 1940s they were joined by the pioneering radiotherapist, Stanford Cade, responsible not only for advances in the treatment of childhood bone cancers, but also for establishing multidisciplinary treatment teams within oncology. As research at Memorial Hospital in New York was demonstrating that childhood cancers were substantially different from those contracted by adults, Williams, Paterson, and Cade were the only British representatives on the international scene.

In Chapter 2 our focus shifts to leukaemia, and we describe how the mechanisms for distributing medical resources nationally – the National Health Service for treatment and the Medical Research Council (MRC) for research – struggled to replicate the research successes being reported in the USA. A flurry of postwar pharmaceutical advances against infectious disease, and the establishment of numerous cancer registries, brought the true magnitude of childhood cancer fatality to the foreground. Although some progress had been made against a number of solid tumours, acute leukaemia was stubbornly unresponsive to surgical and radiotherapeutic intervention. However, pharmaceutical research in the USA was providing hope, with a number of chemical compounds apparently briefly arresting the disease. As the work of the paediatric section of the Atomic Bomb Casualty Commission in Japan was demonstrating the leukamogenic effect of radiation, particularly pronounced in the youngest people exposed, research in Great Britain was revealing increases in the incidence of leukaemia and a new spike in cases in the very young that could not be entirely explained by diagnostic improvements – a situation also evident in other countries. We examine the controversies surrounding research into the connection between radiation and leukaemia, particularly the work of epidemiologist, Alice Stewart.

The development of cooperative groups and large-scale cancer chemotherapy trials is the theme of Chapter 3. Although hugely successful in the USA, for adult and child patients, these processes met with considerable resistance in Great Britain on both ethical and financial grounds. Despite the apparent successes achieved by American hospitals coordinating their protocols and results, physicians in debt-ridden Great Britain were loath to administer such harrowing treatment, particularly to children, for what appeared to be little gain. Many found the idea of random allocation, a central tenet of clinical trial procedure, ethically indefensible. This chapter recounts the attempts in the late 1950s to initiate trials for adults and children with leukaemia, and the problems experienced by participating clinicians.

In Chapter 4 we discuss the work of Denis Burkitt and his colleagues based in and around Kampala, Uganda, in the 1960s, and trace the information networks created between this newly independent country and the cancer centres of the West that led to this rare cancer being pivotal in the history of cancer research. Burkitt's study of the incidence and management of the African lymphoma proved hugely influential in the histories of cancer chemotherapy and studies of cancer causes, providing reasons for optimism as he presented not only charts and tables demonstrating promising results, but also, in an echo of older traditions for the display of medical success stories, parades of healthy saved children. Not only did the tumour Burkitt discovered in 1957 demonstrate the curative potential of the new anticancer drugs, it raised the possibility of a viral cause. The search for a cancer 'germ', often viewed from the vantage of the present as an unfortunate detour and distraction in medical research, has a long and largely hidden history. However, while never the cure-all solution many hoped, it has produced important breakthroughs concerning a number of cancers. For some of these conditions the 'Holy Grail' of cancer vaccination has been achieved therapeutically, and for a few, prophylactically.[1]

We turn to look outside the medical profession in Chapter 5, and begin to consider the wider cultural meanings of childhood cancer in Great Britain as it came to be presented to the public for the first time in the 1960s. The impact of the discoveries of curative therapies on the newsworthiness of children with cancer is traced by analysing some of the earliest coverage of individual sufferers, and of those who tried to help them. With public interest provoked by Burkitt's discoveries and news of others in cancer virus research, the newly founded Leukaemia

Research Fund (now Leukaemia and Lymphoma Research) began to actively court media attention, a strategy frowned upon by the MRC. As public interest was spurred, the relative amounts of state and charity funding into cancer research were increasingly brought into question. With little in the way of effective therapy available in the early 1960s, interest in unorthodox answers flourished. The plight of sick children offered no hope by orthodox cancer treatments was brought forcibly to the attention of the British public in the winter of 1963–64, through the story of a Blackpool family. Four-year-old Edward Burke was the first individual child patient to make the national news when his parents sought to access Gaston Naessen's unproven anticancer serum, Anablast, to treat his acute leukaemia. We consider why this treatment was seen to hold such hope, and how the public and the medical profession were persuaded to relinquish optimism over its potential as news of successful chemotherapeutic protocols developed in American hospitals, shaping newspapers' editorial choices and British government policy. We close this chapter by outlining the history of alternative medicines available to families of children with cancer, and considering the consequences of following the theories proposed and regimes prescribed by alternative therapy providers. A number of families facing bleak prognoses for their children turned to alternative explanations for and responses to cancer, as we show in some detail in Chapter 7 through individual patient stories. It is understandable that disillusionment with dehumanising technology and apparently impersonal medical approaches, serious side-effects, and medical failures, fuelled families' interest in unorthodox approaches, during the decades when for many sufferers the chances of cure were extremely low.

In Chapter 6 we focus on the processes that drove the professionalisation of paediatric oncology from the late 1960s, following the new specialists in children's cancers as they sought support and training, founded clinics, made international connections – primarily in the United States and Europe – and called for higher levels of referral from surgeons and paediatricians. Increasingly in both research and medical care, paediatric malignancies were accepted to be radically different in kind from adult cancers, requiring a different therapeutic approach to tackle their tendency to grow fast and spread early, and specialised wards and clinics where researchers and clinicians could build and wield expertise. Despite opposition, totally centralised care and the recruitment of all children into clinical trials – a call that appears to have been uniquely British – came to be the core principles of paediatric oncology in Great Britain. Drawing on the published reflections of the early paediatric oncologists, and on interviews conducted with clinicians who treated children with cancer during this period, we recreate the ways in which members of this new breed of doctor viewed themselves at the time. Faith in the curability of childhood cancers was a motive force to effect wholesale change in the organisation of care for these children.

The organising strength and research power of the paediatric oncology profession, in Great Britain and internationally, has transformed the survival rates for almost every childhood cancer since the dismal days of the 1940s and 1950s when most patients succumbed within a few months. While some of the rarer conditions remain resistant to all available treatments, many disease types now have such effective therapy plans, evolved over decades of clinical trials, that over three-quarters of those developing the malignancy will be cured with minimal or no apparent long-term side-effects. We close Chapter 6 by mapping the key drivers of steps in medical understanding that led to the nationwide availability of often curative therapies.

In our final two chapters we turn to the experiences of the patients themselves, and the faltering attempts of the medical profession to help them and their families live with uncertainty, with no promise of ever knowing if cure has been achieved. In Chapter 7 we present three patient case studies – from 1960–62, the early 1970s, and the late 1980s – to shed light on the impact that progressive clinical trials and ever-changing survival statistics had on families living through highly experimental times. In this context, we explore the issue of disclosure: whether child cancer patients should be told they are going to die. This contentious debate was effectively brought to a close by the seminal work of Myra Bluebond-Langner who, in a study published in the late 1970s, declared that children were fully aware of their impending fate, but feigned ignorance to protect their loved ones. However, as the medical world absorbed this information, a new challenge was appearing: how to support the increasing numbers of patients surviving their treatment and returning to 'normal' life. We discuss Comaroff and Maguire's influential research conducted with families treated in Manchester in the late 1970s – one of the first studies to investigate this uncertain state.

The last case study in Chapter 7 follows one of the children unsuccessfully treated in Manchester, a decade after Comaroff and Maguire's survey. The memoir of Jimmy Renouf's treatment offers a rich account of the emotional journey followed by families taking part in clinical trials, highlighting the anger that parents can feel at being on the 'wrong' trial arm. Renouf's experience also focuses our attention on the scientific controversy a few years earlier concerning the high numbers of children suffering from cancer near the Windscale nuclear power station (now known as Sellafield). The Windscale excess of cancer cases, and the corresponding government investigation, contain multiple echoes of the 1950s' fears that radiation would poison the innocent.

In the closing chapter, we look at the problems cancer survivors face, and the responsive development of psycho-oncology, a scholarly attempt to understand and meet the needs of those who live past the end of treatment. We examine the problematic history of this speciality, and its early tendency to pathologise the coping strategies employed by survivors and their families. We review the studies that have explored different aspects and impacts resulting from the late effects of treatment and the uncertainty of remission, including the observation that many survivors report a desire for ongoing support and expert advice long after their clinical appointments have come to an end. The first generations of childhood cancer survivors have generally faced odds far worse than those confronted by sufferers today, and lived through more toxic treatments, but the psychological scarring affecting many survivors appears to be a shared result of facing uncertain outcomes. We trace the early development of local provision for this ever increasing group at the major treatment centres for childhood cancer in Great Britain, which often imported American ideas, adapting them to build an expanding native structure.

Thus, the needs of the cured have become a significant area for research and for investment in therapeutic services, as the medical profession seeks to better equip future generations of survivors to truly leave cancer behind. We conclude with the hope that our historical review of what childhood cancer has been will help shape current debates about what childhood cancer now is and can become, and emphasise the enormous transformative effect that even small research groups may have when there is cooperation between medical specialisms, between hospitals, and between countries.

1

Childhood Cancer: A Disease Apart

A boy in his early teens lies on a bed under a 200-kV X-ray machine, preparing to receive his daily dose of radiation, treatment for the metastases spreading from a tumour in his testicles. The radiation is being administered through a large circular applicator, ensuring even distribution over the whole abdomen in the hopes of catching every malignant cell in the region; the area around the applicator is shielded with lead and rubber to protect other tissues. He receives daily blood tests to monitor the effect of the treatment: a depleted number of lymphocytes – a type of white blood cell – indicates the treatment is taking effect, but if the count drops too far treatment will have to stop, as the risk of infection becomes too high. He will also be monitored for radiation sickness, the initial doses he receives being low, then increased until his point of tolerance is reached. The treatment will take about three weeks. During this time, he will be an inpatient in the Christie Hospital in Manchester, receiving his daily radiotherapy at the associated Holt Radium Institute (Figure 1.1).[1]

This boy's chances of survival were as good as they could be anywhere: the large new cancer hospital in Manchester was renowned worldwide for its innovative and successful application of radiotherapy in the treatment of metastatic cancers. The therapy was intended to be curative – that is, the doctors managing his care subjected him to this long and dangerous course of radiation because his kind of tumour was known to respond strongly to radiation and to be, on occasion, totally curable. Unfortunately, the research paper in which his treatment was described, published in 1940, tells us little about the long-term effect of the regime on his general health, his state of mind, or his survival. Nonetheless, his story is an interesting one, for he was one of the first children to be treated this way in a British hospital: one of the first to

9

Figure 1.1   Child receiving radiation at The Christie Hospital in the 1940s
Source: Tod (1940), reproduced by permission of The Christie.

receive state-of-the-art radiotherapy, in a dedicated cancer hospital, where the doses he received were calculated specifically for younger, and therefore smaller and still growing, patients.

Until the 1930s, childhood cancer was almost invisible. The majority of children who developed leukaemia or solid tumours were not diagnosed correctly, and treatment offered to those who were was limited: their early death was taken to be inevitable. Few doctors, even specialist surgeons, would come across more than one case of childhood cancer in their entire careers, so no expertise had been built up in general hospitals. In the 1930s, however, owing to observed increases in mortality rates and systematic attempts to assess and increase the utility of radium as a treatment, all kinds of cancer began to attract more attention. A number of the doctors interested in cancer began to tally the tumours they saw in children, and interested pathologists declared them different in kind from adult malignancies. As increased resources were poured into research on cancers commonly affecting adults, interest in paediatric malignancies, by virtue of their very peculiarity, became a special area of study for a small number of researchers and clinicians. Thus, childhood cancer began its slow transformation from a hidden disease of which a few doctors had little knowledge, to a research hotspot, generously funded by governments and charities, an especially fruitful topic for exploring the mechanisms of cell growth and division.

Some of the interested doctors were haematologists, members of a new specialism dealing with blood diseases, including leukaemia. All haematologists had laboratory facilities, for blood was an ideal – and easily accessible – tissue for investigations; and a few, especially in the USA, also had control of some hospital beds, allowing detailed study of the effects of blood diseases in patients, as well as in laboratory samples. In Chapter 2 we will examine work, following the Second World War, on leukaemia and other blood cancers as they became central to cancer studies, but in the 1930s few connections had been made between the study of blood diseases and research into solid tumours. As solid tumours accounted for the vast majority of cancer mortality in the population as a whole, leukaemia was not seen to be of particular research importance.

The Changing Place of Cancer in the Medical Imagination

Historically, cancer has not been an easy disease to treat, or to be seen to treat. Until the late nineteenth century, almost all patients suffering from the disease died whether they received medical attention or not. Thus, the majority of hospitals explicitly sought to prevent cancer patients from taking up valuable bed space, especially if the hospitals were charitably funded and wanted to appear effective. Institutions for cancer were rare, and, as Pinell (2002) has shown for France, were usually intended to serve as places where cancer patients could die. That such deaths were often painful and very distressing was seen as a valuable test of the devotion of attendant nuns.

In the early nineteenth century, only one of England's charitably funded general hospitals catered for these otherwise shunned patients: the Middlesex Hospital had a cancer ward from 1792. Among the many specialist hospitals that developed in Victorian London, there was one for cancer, the London Cancer Hospital (now the Royal Marsden), from 1851. From the 1880s similar hospitals were opened in Leeds, Liverpool, Manchester, and Glasgow; typically, these fused pity for the dying and a new interest in medical research and possible remedies (Murphy, 1986; Pinell, 2003). From the 1890s, the idea of organised scientific and medical research had taken root in the universities of Great Britain and the USA, following German precedents, with some investigations being funded in the expectation of useful results. Cancer became a popular subject, in part because it was understood to be a disease of cells – the product of excessive cell division – and thus a suitable route to a greater understanding of the basic building block of life.

At about the same time, scientifically minded surgeons were developing ambitious new operations to root out cancer. With anaesthetics available from the middle of the century, the introduction of antisepsis in the 1860s, and aseptic techniques during the 1880s, patients could tolerate more invasive procedures and dramatic surgical interventions became survivable. This extension of surgical procedures was closely linked with the new science of bacteriology, especially in Germany and those British and American hospitals that adopted a scientific approach to surgery. The radical new cancer operations were also influenced by new pathology theories, which understood cancers as local diseases that became generalised as cancer cells moved into other parts of the body: surgeons also attempted to remove the tumour's surrounding tissue in order to capture any cells on the move.[2] The discovery of X-rays in 1895 introduced a new modality of treatment that proved successful for some skin cancers. Soon afterwards came the discovery of radium, a miracle element exploited in small doses as a tonic, which, deployed in larger quantities, could burn away tumours. Around the turn of the century, these fresh intellectual and clinical avenues prompted the creation of many new organisations committed to understanding or combating cancer.

With surgery and medicine becoming more ambitious and funding for research increasing, cancer emerged from the shadows. This focus increased between the wars, along with cancer mortality rates, chiefly owing to ageing populations. The considerable dangers of X-rays and radium had also become clear at that time, as many of the earliest scientists and doctors to work with radiation had fallen sick from its effects. Most countries tried to control the distribution of radium, limiting large supplies to specialist cancer centres where appropriate expertise would ensure safety.[3]

As noted above, the Middlesex in London was the first British hospital to specialise in the care of cancer patients, and many of those instrumental in the founding of Great Britain's first cancer research charity in 1902, the Imperial Cancer Research Fund, were active on its wards. Initially, the supported research was surgical, but it rapidly became clear that even apparently well-contained tumours could not be cured through surgery alone, and by the 1920s reports were arriving from radium centres in continental Europe suggesting that radiation, either alone or in combination with surgery, might be more successful. In the early 1920s, British research into the therapeutic possibilities of radiation, delivered as X-rays or as gamma rays from radium, was taken under government control, with the Middlesex a leading site for experimentation around therapeutic possibilities, especially for tumours where surgery was difficult or ineffective.

Though both expensive and dangerous as a medical treatment, radium appeared to hold great promise and, during the 1930s, several governments began to establish regional centres for radiotherapy. Owing to the pioneering work of Marie Curie and her husband Pierre, France was a leader in this development, with the Swedish centre in Stockholm also becoming an international model. In Great Britain, radium supplies had been collected for medical use in the First World War, and were therefore controlled by the government. The Medical Research Council, established just before the First World War, had realised that radium could link medicine and physical science, thereby stimulating the growth of more scientific methods in clinical research (Murphy, 1986; Pickstone, 2007).

The 1939 Cancer Act recommended the provision of radiotherapy be extended through a network of regional centres for cancer sufferers. Although the Second World War prevented nationwide implementation of the proposals, the Act can be viewed as the expansion of a process already well underway: by the end of the 1930s, the old specialist centres for radiotherapy treatment, in London, Manchester, and Glasgow, had been joined by new regional centres where radiotherapy was hailed as an innovative and properly 'scientific' treatment for cancer, one which might succeed in cases where surgery failed.

In the USA, several cancer hospitals and research centres had established international reputations by the end of the 1930s. The Memorial Hospital for Cancer and Allied Diseases in New York was especially noted for its studies of childhood cancer. The Mayo Clinic, in Rochester, Minnesota, was also considered to be a cutting-edge research centre, for its refinement of diagnostic radiology to detect and measure cancers, and for a separate department researching therapeutic radiotherapy. However, despite these research hotspots American radiotherapy did not become generalised as a separate specialism until long after the Second World War, owing, in part, to the dominance of surgeons in cancer care management and partly because of the medical profession's opposition to any state control of radium.

In general, American conditions of practice encouraged specialisation, as doctors could both practice general medicine for a group of regular patients and undertake specialist work in hospital wards and clinics in a way not easy to manage in Great Britain, even before the National Health Service (NHS); radiotherapy was peculiar in this regard, emerging as a specialism later in the USA than in Europe. Paediatrics, as we shall see, was a favoured 'semispecialism' of those American doctors who saw themselves as catering to families; the majority of child-specific work in Great Britain was undertaken by general practitioners and school-delivered medical services. Specialists in paediatrics were few in number, restricted to children's hospitals in the big cities, where they would manage the care of children suffering from rare conditions.

The Organisation of Paediatrics in Great Britain as a Medical Specialism

Paediatrics, the branch of medicine concerned with the care of children and infants, had become relatively well established in mainland Europe and the USA in the early twentieth century, but did not take shape as an organised specialism in Great Britain until the 1930s. Professional associations were founded to promote the interests of those seeking a career specialising in treating childhood illnesses, including the American Medical Association's Pediatric Section in 1880 and the American Pediatric Society in 1888. In major cities across the USA and Europe, doctors successfully practised as paediatricians, yet in Great Britain very few medical practitioners established themselves as paediatricians until the 1930s (Historical Archives Advisory Committee, 2001). The first historian of the British Paediatric Association (BPA), Hector Charles Cameron, noted the reasons for this in his account of the Association's first 24 years:

> The delay in the appearance of the paediatric specialist was not due to lack of interest in the subject. It was a result of the peculiar organisation of the profession found in this country alone, divided as it is into two unequal parts, a larger of general practitioners, a smaller of consultants (1955, p. 1).

In continental Europe, state institutions that housed children had provided a basis for paediatric specialisation and education, but tax-funded institutions in Great Britain were rarely used as an avenue for studying children's health or treating their diseases; until the establishment of the NHS in 1948, medical education was largely delivered in charity hospitals, which only admitted children in exceptional circumstances. As doctors' work in charity hospitals was unpaid, it was very difficult to establish any specialism that did not also attract a significant volume of private practice; wealthy families with sick children were usually content to consult their regular family physician rather than seek out an unfamiliar specialist.

Partly owing to this exclusion of children from the majority of medical charities' services, larger British cities had supported charity outpatient services that focused on the needs of children. The first of these was created in Holborn, London, in 1769, but lacking sufficient funding it closed its doors after 12 years. A second dispensary for children was launched in London in 1816, and while it languished after the death of its founder in 1824, it was subsequently reinvigorated by Charles West, who went on to create the Hospital for Sick Children in Great Ormond Street in 1852 (Lomax, 1996, pp. 5–6). Manchester's medical provision for children similarly grew from an outpatient service. The General Dispensary for Sick Children started treating patients in 1829; it remained a small charity until the middle of the century, but inpatient treatment was established in 1859, and a decade later work had begun on a purpose-built building in the suburbs of the city, where children could be treated in cleaner air. This was the Royal Manchester Children's Hospital, Pendlebury (Barnes, 2010).

Social and medical interest in the health of mothers and children had been on the increase since the 1850s, with a growing number of maternity hospitals admitting sick babies. Public health campaigns against infectious diseases also provided support for children's hospitals, as did the emigration to Great Britain of a few European-trained doctors with radical political ideas regarding who should have access to healthcare. The expansion of surgery in the 1880s, as noted above, also affected the flow of patients to children's hospitals, increasing the numbers applying for admission and attracting a small number of middle-class families alongside the poor.

Until this time, then, children's hospitals had existed because children were excluded from the general hospitals, not because of any specialised knowledge possessed by their medical staff. However, at the end of the nineteenth century, as a new focus on 'child studies' developed, this situation began to change. Doctors and scientists became increasingly interested in diseases of development: as these were so unlike adult diseases, paediatricians acquired a distinctive intellectual claim to expertise.

Interest in children was then greatly enhanced by the British government, which, like most others, began to promote the health of the population in order to secure the strength of the nation: for a nation to successfully compete in an age of industrial and colonial rivalry, it needed a large and healthy population. Welfare systems for maternity, infants, and schoolchildren became a national priority, in Great Britain and across Europe and the USA (Cooter, 1992). Initial interest focused primarily on infectious diseases, but following the First World War, researchers and clinicians in Great Britain, the USA, and Europe widened the focus to include children's development, studying infant feeding and family diet, and researching the biochemistry behind health and disease in efforts to explain the widespread debilitating health conditions observed in young men called up for military service (Shulman, 2004). As both interest and services increased, British doctors concerned with child health began to establish professional organisations. This enabled them to call for greater investment in funded hospital posts in paediatrics, where the study of child *health* could be furthered alongside the provision of treatment for childhood *illness*.[4]

Dining clubs for children's doctors were meeting regularly in London from 1918, in Scotland from 1922, and in the provinces from 1924 (Cameron, 1955, pp. 3–6). Canadian paediatrician, Donald Paterson, who had moved to Great Britain after the First World War and tirelessly advocated for a national professional body on the model of the American Pediatric Society, brought these informal societies together. The inaugural meeting of the BPA, with Paterson as honorary secretary, was held in 1928. That same year, Birmingham Children's Hospital became the second hospital in the country to fund a Chair in child health, awarding the post to Sir Leonard Parsons, who had trained at Great Ormond Street Hospital in London (Stevens, 2001; Dunn, 2006).[5]

Liverpool, Glasgow, Edinburgh, and Sheffield quickly followed in founding chairs in child health; Norman Capon, Geoffrey Fleming, Charles McNeil, and Albert Naish each took the title of professor, and each served as president of the BPA for a term (*BMJ*, 1952; Jackson, 1988). Thus, in the 1930s there was scope for medical students to opt for courses and placements in paediatric departments, and to join an organisation committed to sharing the best research into the causes of and treatments for the full spectrum of childhood ailments. It was now possible for a doctor or medical researcher to develop a career studying only the diseases of childhood (Valman, 2000). However, until the foundation of the NHS, paediatrics remained an extremely small specialism, chiefly restricted to the charitable children's hospitals in the main cities. Those doctors who dealt with children in general practice, in school clinics, and in infectious disease hospitals or welfare institutions were not considered to be specialists.

The Beginnings of Research Into Childhood Cancer

The recorded incidence of cancer in childhood steadily increased from the 1930s onwards, as more cases of children with suspicious lumps or unusual levels of cells in their blood were referred to the rare hospitals with known specialisms in childhood cancer. Some hospitals developed dedicated wards and clinics for children with leukaemia and solid tumours, and supported physicians' attempts to initiate tests of new therapies, comparing results over time and across institutions. To facilitate further research, a small number of cancer hospitals and large teaching hospitals began registries of childhood cancer cases in the early 1950s. Pathologists in these places, by seeking out samples of malignant growths removed by surgeons in other hospitals, expanded their records of the incidence of each kind of tumour across entire regions. These changes in treatment and investigation had their roots in the studies undertaken by a handful of curious researchers during the 1930s, studies that established the distinctiveness of children's cancers by asking what differentiated children's tumours from those of adults.

Appointments of specialists to manage the hospital care of children with cancer, and to examine the cancer cells taken from child patients, were first made in the 1930s, but were exceedingly rare, especially in Great Britain. The proportion of children with cancer seen in clinics or wards run by these specialists was minute. Most parents could not afford or access cancer hospitals and, given how rarely the condition presented in the consulting room, few children's hospitals employed a doctor devoted to cancer. By the end of the decade, however, the handful of people studying childhood cancer were beginning to interact and develop their expertise, publishing on the subject and travelling to one another's centres. For British doctors and researchers, this meant visiting American cancer hospitals investing in studies on the malignancies that particularly affected children.

In the USA, training programmes in paediatrics had been established at most medical schools by the end of the 1920s. The influence of paediatricians on the funding priorities of medical schools, as well as on state and federal health policy, increased over the following decade. The American Academy of Pediatrics was founded in 1930, followed a year later by the Society for Pediatric Research for academics studying childhood diseases outside clinical practice. Subspecialisms flourished in the largest medical schools on the back of research success in cardiology, endocrinology, and haematology (Historical Archives Advisory Committee, 2001; Shulman, 2004), and some graduates of the expansive paediatric training programmes specialised in the study of tumours.

In 1937, Sidney Farber, who later became famous for securing the first remissions of acute leukaemia, published what became the standard textbook of paediatric pathology, *The Postmortem Exam*, the product of a decade's experience examining tissues taken from children who had died from illness. Farber had taken the position of resident pathologist at Boston's Children's Hospital in 1928, where his duties included conducting autopsies on patients and diagnosing malignancies: learning to recognise the signs of cancer in the bodies and cells of children. At the same time, James Ewing was assembling a team of specialists to study tumours in children at Memorial Hospital in New York (Krueger, 2008). Papers published by this team in paediatrics journals through the late 1930s noted that the distribution of cancer by type in children was very different from that observed in the adult population. The Memorial research team also documented response rates from surgery and/or radiotherapy at odds with what one would expect if cancer behaved the same way in children as it did in adults.

These New York studies discovered that sarcomas were far more prevalent in children than carcinomas, a reversal of the situation for adults. They also found there were many tumours relatively common in childhood that were never seen in adulthood: Wilms' tumour (a type of kidney cancer) and retinoblastoma (cancer in the eye) typically appeared in the very young, and in some cases appeared to be heritable, presenting possibilities for additional research into cancer genetics. Neuroblastoma was another tumour only seen in very young children, the result of embryonal cells failing to mature and instead replicating rapidly and metastasising widely; yet some children with this tumour experienced spontaneous and complete remission, cells abruptly ceasing proliferation and switching to their mature form, suggesting exciting lines of research into what induced cancer cells to replicate ceaselessly, rather than obey the usual rules of cell division. Cancers common in adults – those of the stomach, breast, and lung, for instance – were almost never seen in children, indicating that some cancers were probably brought on by occupation, diet, or poisoning: by damage accumulated over a lifetime. That these cancers were nonetheless occasionally reported in children also aroused interest, for the chance such cases afforded for examinations of medical and family histories, looking for possible environmental triggers of malignant growth.

Research into how to recognise, classify, and treat childhood cancer in the 1930s was, therefore, widely dispersed but conducted on a small scale. The vast majority of potential patients slipped through the loosely strung net of centres collecting details on leukaemia and the tumours of childhood. But given that the work was seen as important for *all* cancer research and treatment, journals were quick to publish findings. In 1940, the papers published by Ewing's team at Memorial Hospital were reissued in a volume entitled *Cancer in Childhood*, edited by Harold Dargeon, the paediatrician in charge of the hospital's children's ward. The object of the volume was to establish cancer as one of the important diseases of childhood. Its publication was charitably supported, enabling free distribution to all medical school libraries in the USA; copies were also sent to the major medical training centres in Great Britain. Many of those doctors who later set up their own research groups proudly declared that they had borrowed, or even come to own, copies of this text during their paediatric training: familiarity with the work of the Memorial team would come to be a stamp of pedigree amongst paediatric oncologists (Taylor, 1990).

Cancer in Childhood became the standard text on paediatric tumours on both sides of the Atlantic, remaining so through the 1940s and 1950s, when there were few publications available to consult. It offered clinical guidance on how best to combine surgery and radiotherapy for cure or, more likely, palliation. Ewing's introductory essay noted that 'As a rule malignant tumors of infants and children progress rapidly and metastasize widely, recurrence is prompt, and the mortality is high' (Dargeon, 1940, p. 17). Echoing the medical profession's line on adult cancers at that time, specialist clinicians held that only those tumours detected early could be successfully combated, with a combination of surgery and radiotherapy. However, many children with apparently early-stage cancer suffered relapse, suggesting that cells in paediatric tumours divided and spread more rapidly than those in adult tumours.

News of the work being carried out in New York, which could boast several world-class cancer research establishments, had spread to Europe even before the publication of this collected volume. In Great Britain, as we have seen, paediatrics was not well established as a separate specialty; there was no requirement for physicians intending to specialise in the care of children to take examinations in the subject area as part of their qualification. *Sub*specialism within the field of child health was even less likely; yet a few paediatric trainees did see ways to combine specialist modules of study within paediatrics and from other areas of practice. One such trainee, Ivor G. Williams, found a way to fuse his three main interests: the care of children in hospital, paediatric surgery, and radiotherapy.

Williams had qualified at the Middlesex Hospital in London and initially planned to specialise in paediatric surgery, but, swayed by his time in the radiotherapy department, took a position as a Rockefeller Fellow in 1938, based at the Crocker Institute of Cancer Research at Columbia University in New York. The directors of the Crocker Institute, worldrenowned for research into different means of administering radiotherapy, had assisted in the development of the machinery required for deep X-ray therapy, researched by Ernest Lawrence and David Sloan, which Williams had seen in use back at the Middlesex. Williams made weekly pilgrimages to Memorial, attending Dargeon's ward rounds and clinical meetings, and Ewing's weekly classes on the management of children's tumours. His earlier interest in using radiotherapy to treat benign growths in children expanded into a study of its potential for controlling malignant tumours.

When Williams returned to Great Britain he published on the new super-high voltage machines he had seen used in New York hospitals, comparing the effects of machines with powers in excess of 1,000 kV to those operating at 250 kV, commonly used in British radiotherapy departments (Williams, 1942). Williams was impressed by the precision with which the high-power beam could be directed, minimising damage not only to surrounding tissues and the skin at the point of beam entry, but also to bone marrow; patients responded to the radiation more quickly and with fewer side-effects. Although no mention was made of children in the published review of his trip, Williams was to use his New York experience in his first senior appointment to wield this clinical knowledge when treating children with deep-seated tumours.

During the war, while treating cancer patients in the Middlesex's relocated radiotherapy department in Mount Vernon Hospital, Williams began working towards the Fellowship of the Faculty of Radiology, which had as one requirement the production of a written thesis.[6] Williams decided to undertake a survey of childhood cancer in Great Britain. Working through records of all childhood tumours and cases of leukaemia from three London hospitals – the Middlesex, the North Middlesex County, and Mount Vernon – he covered the period

1927–42; however, finding such cases proved hard work owing to poor hospital records prior to 1931. Noting that the cases he found could not be taken as a reliable measure of incidence, if only because of the problems with record-keeping, he reviewed previous analyses by other authors who had attempted to measure the incidence of cancer in the young. The very scarcity of such studies – his review discovered only four, dating from 1876, 1883, 1931, and 1940 – confirmed the novelty of researching how cancer struck children.

A summary of his thesis was published in the *British Journal of Radiology* in 1946. The survey reported the clinical course of the more frequently seen paediatric tumours, and recounted how rarely surgery and radiotherapy could prevent disease from progressing. Many tumours found in children had already grown through and around major organs and blood vessels, making their surgical removal extremely hazardous if not impossible. Williams therefore concluded that 'Many tumours are biologically inoperable even if surgical removal is feasible anatomically' (1946, p. 193). Radiotherapy was also of limited use in children as it is based on the principle that radiation kills the rapidly dividing cells of a tumour while leaving the surrounding tissue, where cells are dividing more slowly, almost undamaged. In children the tissues within which the tumours develop are themselves growing rapidly, making it exceedingly hard to devise treatment that can affect a tumour while leaving healthy tissue unaffected.

Williams devoted a major part of his professional career to improving radiotherapy treatments for children's cancers. In 1948, he became the head of the Department of Radiology at St Bartholomew's Hospital, where he supervised research into the use of new mega-voltage X-ray machines – the hospital was equipped with the only million-volt apparatus in the country – and made paediatric radiotherapy a specialty in its own right. St Bartholomew's Hospital – established in 1123 and occupying the same site for almost 900 years – had some of the country's best facilities for the treatment of cancer and a long-standing reputation for being a leading institution in the development of innovative forms of medical education and training (Waddington, 2003).[7] Williams' position therefore was a prestigious one, and an acknowledgement of his status as an innovator and expert in radiotherapy. He managed the radiotherapy service for children referred to the hospital from other hospitals across London that lacked facilities until the early 1960s (Figure 1.2).

Paediatric radiotherapy was also being developed as a special area of practice in Manchester, at the Christie Hospital and Holt Radium Institute. These two institutions were united in 1932 under a new

Figure 1.2   Ivor G. Williams with a young patient
Source: Courtesy of Professor Jim Malpas, by permission of St Bartholomew's Hospital Archives.

director, Ralston Paterson, a leading authority on the treatment of cancer by radiotherapy who had trained in the USA. Under his leadership, the Christie Hospital pioneered routines that were taken as a gold standard by many other institutions across the world. He and his wife, Dr Edith Irvine-Jones Paterson, also established a programme of basic research into cancer, particularly in the field of radiation science.

Edith Isobel Irvine Myfanny Jones had trained at Edinburgh's medical school, along with her future husband. She then held a post as resident paediatrician in San Francisco's Children's Hospital and took up a scholarship for further paediatric study at the Washington University School of Medicine in St Louis. There she was able to combine clinical duties with laboratory studies, and tour the major paediatric centres in the USA, gaining exposure to levels of clinical specialisation not available elsewhere. Edith and Ralston married in Manhattan in 1930 shortly before moving back to Scotland. She relocated to Manchester when her husband was appointed director of the newly amalgamated cancer centre the following year. Having never previously worked on cancer research or treatment, she embarked on self-directed training in cancer therapeutics, and was formally employed by the Christie Hospital in 1933 to conduct research on tissue culture. Here she initiated a programme of experimental radiobiology. She began by working alone and with no budget, but gradually secured funding, laboratory resources, and technicians, pushing her programme forward. By 1938, she was publishing papers on her research exploring the effects of different radiation doses and fractionations – ways of spreading a therapeutic dose over time to allow healthy cells to recover and so reduce side-effects – on tumours experimentally induced in specially bred mice, work on biological dosimetry that earned her international renown (*BMJ*, 1995).

While her published work centred on mouse experiments, Edith Paterson's research was always directed at improving outcomes for those suffering from paediatric malignancies. She used mice to explore new possibilities that, if promising, were translated into the clinical care of children with tumours. She established a children's ward within the Christie building, the first of its kind in any hospital offering cancer treatment to children, equipped with toys and staffed by nurses with an interest in caring for children. Edith Paterson, like Williams, was by initial training a paediatrician, and thus sought to meet the needs of the child, as well as those of the cancer.

Manchester and London each boasted good working relationships between a children's hospital, where paediatric research was supported, and a hospital specialising in the use of radiation treatments for cancer to supplement surgery. There was in each city, therefore, scope to integrate research into paediatric pathology with clinical evaluation of response to different combinations of treatment options.

After the Second World War the status of Great Ormond Street Hospital for Children as an international research centre in paediatrics was confirmed and enhanced by the creation of London University's Institute of Child Health, a postgraduate teaching and research centre, under the professorship of Sir Alan Moncrieff. Many London children diagnosed with tumours in the abdomen, in soft tissue, or the brain were referred to Great Ormond Street Hospital, where paediatric surgeons, who specialised by site of disease rather than disease type, managed their care. If radiotherapy was deemed necessary, patients would be sent across London to St Bartholomew's for postoperative treatment. Treatment for and research into childhood tumours in this cluster of specialist hospitals remained largely under the control of surgeons, as it was they who planned patient care.

In Manchester, The Royal Children's Hospital at Pendlebury on the outskirts of Manchester was the paediatric training hospital for the medical school of Manchester University, with Professor Wilfrid F. Gaisford as the new head of paediatrics. Patients were sent to the Christie for radiotherapy, where research into paediatric radiotherapy was developing an international reputation. Gaisford was keen to support the growth of the city's reputation as a centre of excellence, working for longer survivals and better care for young patients and their families.[8] His ambitions to offer a regional referral service chimed with the cancer service that Ralston Paterson had built with general hospitals across the region, where patients were seen by Christie staff in their home towns and referred to the central hospital when appropriate. For children with cancer, staff from Pendlebury and the Christie worked hard to ensure that all cases in the region found their way to the specialised facilities available in Manchester; in turn, this concentration supported research into paediatric malignancies.

In the 1940s, London and Manchester were the only two cities in Great Britain to offer fully integrated care for children with cancer. Yet Edith Paterson and Ivor Williams are little known outside the profession of paediatric oncology for their pioneering research into childhood cancer. Paterson has been the subject of historical study on account of her early chemotherapeutic research, screening chemicals manufactured by ICI for antitumour properties, and Williams is better known to historians for his research and writings on the use of radiotherapy in the management of breast cancer, and as a leading light in the radiology profession.[9] We would argue that the most significant aspect of the careers of these two doctors was their demonstration that bringing children with tumours under the care of medical teams specialising in cancer treatments *and* paediatrics yielded both better outcomes and productive research. They proved the value, professionally and medically, of seeing childhood cancer as a disease apart.

Treatment for Tumours in the 1940s

British research into paediatric tumours in the 1940s primarily sought to improve survival rates through changes in surgical technique, and through better targeting of radiation to kill those cells which could not be removed. During this period, children with solid tumours received more efficacious interventions than did those with leukaemia (which we will discuss in Chapter 2) and it was with solid cancers that experimentation was making headway, improving survival rates and reducing the complications and costs borne by those who lived. While surgery became increasingly less hazardous through improvements in antisepsis and anaesthesia, radiotherapists calculated optimal doses and devised fractionations for treating small bodies and designed tables that could hold young, sedated children still during treatment. Survival times were improved for patients with bone tumours by giving radiotherapy after amputation, reducing the risk of local recurrence. Infants with Wilms' tumour still contained within the kidney could in many cases be cured through postoperative radiation to the abdomen.[10] Many children with localised Hodgkin's disease, a type of lymphoma, were cured through the use of radiotherapy.[11] To minimise the risk of excessive bone marrow suppression and of permanent damage to growing tissues, radiotherapists learned how to split doses; and they monitored blood counts closely, in recognition that the bone marrow of children was more easily suppressed by radiation than that of adults. In each of these cases, the discoveries and new systems of treatment that effected change originated from research hospitals where individuals worked in interdisciplinary teams, and had good relations with the house medical staff and nurses.

Williams' review of 181 children treated for cancer in London in the late 1920s and the 1930s recorded the emergence of standard forms of treatment for the different types of malignancy then recognised. Skin carcinoma might be treated with X-ray therapy, or even with radium, following surgery to remove all obvious cancerous growth. Patients with abdominal or brain tumours were given deep X-ray therapy in most cases, usually, if surgical removal was possible, postoperatively. Children with 'lymphomsarcoma' (lymphoma in modern terminology) were given high doses of radiation over broad areas of their bodies, designed to arrest cell division in the affected glands. 'Glioma of the retina' (now known as retinoblastoma) was tackled through excision of the worst affected eye, and the insertion of radium seeds in the other. Bone tumours, following surgical removal when possible, were subjected to deep X-ray therapy in an attempt to halt the almost ubiquitous metastases in the lungs.

Both radium and deep X-ray therapy were expensive to set up and to run, and only available in specialist hospitals, but for those children who did receive state-of-the-art therapy, there was some hope: roughly one-third survived their cancer for at least a few years. Five-year survival, however, was closer to ten per cent, and some tumours were far less tractable than others, especially those of the bone. Tables of treatments and outcomes demonstrate that for many patients cancer returned, often in sites far from the primary tumour, and that no available technology could satisfactorily deal with such recurrence. Metastasis, at time of diagnosis or subsequently, carried a dire prognosis.

The treatments given in Manchester in this period were also documented in published reviews of cases seen over long time spans stretching back to 1933. The paediatric radiotherapy service at the Christie and

Holt saw a large number of children with lymphoma, or with tumours in bone, brain, or abdomen – the other childhood malignancies occasionally being referred through for specialist treatment. The radiotherapists there, guided by Edith Paterson's research, secured remarkably high three- or five-year survival figures of 35 per cent for children with brain tumours, 20 per cent for infants with neuroblastoma, and 30 per cent for children with Wilms' tumours.[12] The doses required, however, did considerable and permanent damage to surrounding healthy tissue, affecting bone growth, hormonal development, and brain function, all of concern to the hospital's growing paediatric radiotherapy team.

The Christie Hospital and Holt Radium Institute employed a number of academically strong radiotherapists who researched ways of treating the metastases so often seen in children's cancers. Margaret Tod (1940) described the hospital's evolving system of trying to control metastases through the use of broad fields of radiation, delivered in small daily doses to minimise harm done to bone marrow and blood production. Inez ApThomas (1942) detailed a decade of development of techniques for applying radium to susceptible tumours in the youngest patients, such as the innovation of embedding radium seeds into simple sticking plasters to make their application less remarkable and uncomfortable for young children. Publications such as these from the Christie include photographs of young patients lying underneath radiotherapy machines, or suggestions of how mothers and therapists could keep young children still for treatment.[13] Such glimpses into practice reveal much about the ways in which the physiological and emotional needs of children were recognised in this specialist cancer hospital. Radiotherapists shared not only their dosage schedules, but also their techniques for calming patients, and for safely and gently restraining them during each session.

The forms of treatment developed for children with cancer in the 1930s and 1940s were not swiftly replaced with the advent of anticancer drugs in subsequent decades; they persisted for many cancers when new chemotherapeutic agents were no more effective than surgery and radiotherapy, and in regional hospitals where medical staff did not have access to the latest drug trials and expensive nursing facilities that these required.

Paediatric Cancer Surgery

During the 1940s, very few surgeons could build practices specialising only in paediatrics; children's bodies were not deemed different *enough* from adults' to warrant a distinct training and profession. Not until the 1970s did most large general hospitals and teaching hospitals have consultant surgeons responsible for all and only operations on children.[14] Before then, those general hospitals that admitted child patients relied upon surgeons who were specialised by site of procedure rather than age of patient.

However, in those urban areas that developed patterns of referral for unusual cases of childhood diseases, including London, Manchester, Glasgow, Newcastle, and Birmingham, those surgeons who treated all children with disease in a particular organ became more experienced than other surgeons in dealing with cancerous growth. For example, while the average kidney surgeon would only see one child with a Wilms' tumour in the course of a career, in hospitals attached to those universities that had established Chairs in child health, a kidney specialist might see ten or more such cases over the course of a decade. These practitioners, affiliated to or teaching within medical schools, occasionally recorded the literature reviews they conducted on tumour removal, along with any innovations they made to technique.

Paediatric surgeons and radiotherapists were not alone in these attempts to step up survival rates. Stanford Cade was perhaps the most influential contributor to radiotherapy experimentation in the 1940s and 1950s. His ideas on how to manage bone tumours, including Ewing sarcoma and osteosarcoma, both of which affect children and teenagers, quickly won advocates across Europe and North America. Osteosarcoma was known to be highly radiosensitive, but almost all children suffered local recurrence of the tumour after just a few months, or aggressive metastases, typically in the lungs. Cade's approach was to see the tumour as systemic – spread throughout the body's organs and systems – from the moment of diagnosis. He advised his staff at the Westminster Hospital, and others charged with treating bone cancer, to undertake irradiation of the affected limb as primary treatment, meaning only patients still alive 18 months later needed to undergo the trauma of amputation.[15] His intention was to reduce surgical trauma – to avoid making patients who were unlikely to survive have to face losing a limb – and to increase surgical efficacy, so ensuring operations were only performed if and when they were most likely to succeed.

Cade's chief innovation, however, according to those with whom he worked, was his insistence that the person in charge of treating a patient with cancer must not bring to a consultation only his or her special interests and training. At a time when, as he asserted, most surgeons knew nothing about radiotherapy and most radiotherapists knew nothing about surgery, cancer treatment seemed to be a lottery: what a patient received depended on who they first saw. Cade initiated multidisciplinary meetings, where all those charged with patient care met to plan an integrated campaign against each tumour. The emphasis on a team approach to cancer care is a recurrent theme in the history of childhood cancer.

Pioneers and Their Impact

Cade, like Edith Paterson and Ivor Williams, is not best known for his work on children's cancer, perhaps because incremental improvements won through what now appears to be an old-fashioned treatment approach do not seem noteworthy in comparison to the sudden leaps in survival effected by the later introduction of chemotherapy as additional therapy. Yet Cade, Paterson, and Williams were the mid-century radiotherapists whose work has been recognised by their own peers as the most significant in changing care for children with cancer. Each developed treatment regimes that markedly improved the length and quality of life of their patients and other children treated in accord with their recommendations. Each was held up as a pioneer by the next generation of specialists in childhood cancer (Taylor, 1990). The importance of these pioneers, and of the very notion of the pioneer, can be seen not just in published and preserved manuscripts penned by paediatric oncologists, but also in the planning of clinical conferences and courses on childhood cancer, staged from the late 1950s onwards. Key international meetings were organised at Memorial Hospital in New York, the M.D. Anderson Hospital in Houston, the National Cancer Institute at Bethesda Maryland, and at the offices of the MRC in London; Stanford Cade, Edith Paterson, and Ivor Williams were frequent – and the only – British invitees to these congresses.

The history of the development of a special interest in children's cancers in Great Britain reveals that the roots of what is now known as paediatric oncology lie in peculiarly British soil. That radiotherapy was becoming established as a full specialty is crucial in explaining the shape of paediatric cancer care in Great Britain. While in the USA, haematologists and pharmaceutical companies have been central to the story of changing outcomes for children with cancer, in Great Britain, specialists in blood diseases and blood testing have not dominated the development of services or protocols. Instead, paediatric cancer services have developed through the efforts of people primarily interested in *paediatrics.* Pathologists and radiotherapists drawn to the different patterns of development of tumours that affected children founded registries and clinics, which remained key sites for research and training alongside the centres administering experimental chemotherapy for leukemic children in subsequent decades. The history of paediatric oncology in Great Britain is thus particularly worthy of study, for it has emerged from research interests and networks that are rather different from those well documented in histories of the growth of oncology in the USA.

Overall cure rates for childhood cancer remained low through the 1940s and 1950s, held so by the prevalence of systemic disease. Many children presented with widespread disease – tumours that had metastasised, or dispersed leukaemia – which could not be eradicated through surgery or radiotherapy. Treatment for secondary tumours grew less and less effective with each regrowth: organs could not be further cut; tissues could not be given more radiation without the creation of new tumours. Treatment for acute leukaemia offered only palliation: antibiotics to control infection and blood transfusions to alleviate anaemia. It is to leukaemia that we now turn.

2
The Rise of Childhood Leukaemia

The Second World War heralded a dramatic increase in the resources poured into medical services and medical research in both the USA and Great Britain. A new political will and enthusiasm for public investment united with a widespread optimism that medicine could banish sickness, an optimism fuelled by successive medical breakthroughs.[1] The use of Salvarsan to treat syphilis in the 1910s, insulin for diabetes in the 1920s, and the antibacterial sulfonamide drugs in the 1930s, had raised public expectations of medical research.[2] The wartime success of penicillin and the subsequent discovery of other antibiotics led many to believe that infectious diseases would all be conquered.

Research funding was especially expanded in the USA, with chemical therapies central to the research programme: funding proposals concerning childhood cancers were particularly successful. In the two decades following the Second World War, the number of people working in healthcare and research tripled, expenditure more than quadrupled, and federal grants for research started to flow. The American government had established the National Institutes of Health (NIH) in 1930 and funded the building of a complex of laboratories eight years later in Bethesda, Maryland, just outside the capital, Washington. In 1937, the National Cancer Institute (NCI) was created under the governance of the NIH and a law passed permitting the awarding of grants to researchers working outside the government's own facilities (Patterson, 1987, Ch. 5). Although the amount of money initially available for this funding was small, wartime medical research needs and successes prompted a dramatic increase, with the budget for the NIH as a whole growing from $180,000 in 1945 to $4 million two years later, as the Institute assumed responsibility for coordinating all public health and cancer research across the country; after a further three years its budget had swollen to $46 million. This leap in federal support for health research can be largely attributed to the lobbying efforts of the American Cancer Society (previously the Society for the Control of Cancer) as, with mortality rates rising, cancer came to be viewed as a major threat to health (Starr, 1982, Book 2, Ch. 3). The Society ran an ambitious campaign on radio, billboards, and through town hall talks, highlighting the life-saving breakthroughs achieved against bacterial infections and polio, and calling on members of the public to pressure their government representatives to vastly increase research funding, the clear expectation being that cancer would yield in the same way infections and polio had (Patterson, 1987, Ch. 7). Thus, Congress and Senate passed bills pumping significant amounts of federal funds into cancer research, to be administered through the NCI.

In Great Britain there was also expansion in the amount of money made available from national funds for higher education and medical research. As part of the push for national efficiency, discussed in the previous chapter, the British government had passed the National Insurance Act in 1911, which granted health insurance for workers, limited sick pay, and covered the costs of consulting a doctor, but did not provide for dependents.[3] Wartime enthusiasm for delivering a truly universal and socialised system of medical care was challenged by professional bodies and charitably funded hospitals, which objected to every proposal the government put forward. The election of a Labour government in 1945 led to a new round of planning and negotiations, and, in July 1948, all hospital services came under state control.[4]

The Medical Research Committee and Advisory Council had been created in 1913, primarily to contend with tuberculosis; it was reconstituted as the Medical Research Council (MRC) in 1919, when both its remit and resources expanded. The government retained direct control over research into public health through the Ministry of Health, while the MRC assumed the role of decision maker for all basic and biomedical science research initiatives. In the fraught discussions over how the National Health Service (NHS) should be organised, one of the fiercest battles concerned the future funding of medical research. A decision was reached that government-funded clinical research should be directed by the MRC, while the NHS focused on delivery of care. Resources ceded to the NHS for research in teaching hospitals were small in comparison with those available in the research units receiving MRC grants, but teaching hospitals were allowed to keep their amassed endowment funds to support research as they saw fit.[5]

The teaching hospitals were not the only bodies with independent funding to fuel research work: medical charities were also offering resources to researchers and hospital doctors, often to support projects that lay outside MRC interests. In Great Britain, as in the USA, public interest and anxiety about cancer had been rising through the first half of the twentieth century (Toon, 2007, pp. 120–2, 137; Cantor, 2008). The Imperial Cancer Research Fund (ICRF) had been collecting donations for fundamental laboratory-based research into cancer's causes and mechanisms since the beginning of the century, but in 1923 a charity was formed to support clinical cancer research, the British Empire Cancer Campaign (BECC). This body was initially held in suspicion by both the MRC and the ICRF, the fear being that the research programme followed by British scientists and cancer specialists might diverge (Austoker, 1988; Edwards, 2007). MRC leaders manoeuvred carefully in their conversations with the BECC and other alternative sponsors for medical research, attempting to secure agreement over what the agenda, and the methods, of investigation should be. But it was the machinery of randomised controlled trials (RCTs), and the level of postwar funding at the MRC's disposal, that ultimately afforded the organisation control over the nation's clinical research.[6]

Following the foundation of the NHS, an initial task for the MRC was to regulate access to, and use of, the latest wonder drugs, antibiotics. To test the efficacy of antibiotics, they employed RCTs, helping to establish this new practice as the standard for therapeutic research.[7] Many children benefitted from antibiotics, and although young patients were not often included in RCTs, there was much discussion between clinicians about best practice (Wilson et al., 1949; Gale et al., 1952; Dobbs, 1953).[8] Antibiotics were specifically tested in children against infections, such as rheumatic fever, that could prove fatal or lead to serious complications, as the MRC strove to instil order into the patterns of antibiotic use in hospitals and assess the relative value of the various antibiotics then available, against one another and compared with sulfonamides.[9]

The published results of these RCTs document the struggles encountered in children's wards, and record how clinicians in different institutions varied in their approach to local problems. While some found it ethically unacceptable to withhold treatment from patients assigned to the control arm, others found parents in their area disproportionately unhappy about taking part in anything billed as a trial. A few discovered that random allocation to the various arms of a trial came to naught owing to the trial's designers not appreciating the significance of every presenting symptom of an infection.[10] Yet all the published results were in agreement on one point: antibiotics were a valuable weapon in the fight against infectious disease in the most vulnerable child patients: the very young and the already ill.

As deaths from infections declined, so other conditions came to the fore; included in these were childhood leukaemia and childhood cancer in general.

Registering the Scale of the Childhood Cancer Problem

As we saw in Chapter 1, in the 1940s childhood cancer was thought to be incredibly rare. There are many reasons for this family of diseases becoming more visible, one of which was that leukaemia – one of the most common forms of childhood cancer – came to be categorised *as* a cancer. In the early years of the twentieth century, leukaemia would most certainly not have been described as a cancer by researchers or clinicians. Recognised as a fatal disease of the blood, leukaemia was believed to spread throughout the body from its first inception and was thus seen as fundamentally different from solid cancers, which develop around a single point (Piller, 2001).

Leukaemia was understood to be a totally intractable disease, with sufferers not seeming to benefit from rest or changes in diet. Attempts at blood transfusion and trials of X-rays made no difference to disease progression or survival times. The chronic form was well known from the middle of the nineteenth century. Cases of an acute form were diagnosed from the 1890s but these were aberrant and often disputed, and patients died rapidly after admission to hospital. Acute leukaemia was originally believed to be the result of some sort of rare, possibly contagious, infection, but with so few cases reported in the literature its causation and natural history were difficult to establish.

In the early 1900s, a number of researchers studying cancers of the lymphatic system argued that leukaemia bore a remarkable resemblance to the lymphomas. However, leukaemia had no local focus, and it was not clear how a dispersed cancer could originate. In 1917, Gordon Ward, a specialist in childhood diseases, undertook a major epidemiological survey of all reported cases of acute leukaemia, collecting details on over 1400 cases. Ward demonstrated that a hereditary or infectious component was unlikely (ibid, p. 288).[11] Leukaemia's cancerous nature was finally demonstrated in the 1930s, when experimental work revealed leukaemia arose in laboratory rodents exposed to known carcinogens such as tar. Advances in cell staining techniques allowed the cellular processes to be witnessed by laboratory workers, demonstrating leukaemia's proliferative and therefore *cancerous* nature as a defining feature (Goldman and Gordon, 2003, p. 4).

However, the status of leukaemia as a cancer was not settled quickly or easily. During the 1930s and 1940s, the disease could be argued to be either a blood disease that belonged in the domain of haematology, or a cancer requiring the attention of cancer researchers and experimentalists (Wailoo, 1997, pp. 169–70). As almost nothing could be done for patients with leukaemia it had not been the most attractive form of cancer for research prior to the Second World War: as we have seen, those clinicians who wished to innovate in childhood cancer treatment mostly focused their efforts on the solid tumours, where survival rates were improved by using powerful new radiotherapy tools to support surgical interventions; meanwhile, acute leukaemia seemed hopeless, with death following diagnosis within a few weeks.[12]

In the ten years after the Second World War, however, acute leukaemia took on a new importance for the medical profession, for three reasons. Firstly, the number of cases of acute leukaemia in children was seen to rise substantially, as revealed by new cancer registers. The number of children diagnosed with cancer in all its forms was also going up year on year during the postwar years, but the surge in leukaemic children was even greater, the increase being demonstrably higher than improvements in diagnosis could account for: it reflected a true increase in the number of children falling ill.

Secondly, the dropping of two atomic bombs in Japan in August 1945 exposed many thousands of children to ionising radiation, and as radiation's effects on growth and development were studied it became clear that exposure could cause leukaemia. The scale of the risk was not known, yet both the USA and Great Britain were committed to atomic energy programmes, as relations between Russia and the USA and Western Europe deteriorated into a Cold War.

Thirdly, American studies of chemical agents against cancers in adults and children were revealing that some compounds had the potential to halt the progress of both solid tumours and leukaemia. The pharmaceutical industry appeared to hold the key to curing those cancers that resisted surgery and radiotherapy. This chapter will explain how these three developments brought the attention of the medical profession onto childhood leukaemia.

Cancer Registration

Whenever a dramatic increase in incidence occurs, one has to ask whether the change reflects actual patterns of disease or changing patterns in diagnosis. With infectious diseases, variation over time is expected, as waves of contagion move through populations more or less immune. Illnesses brought on by exposure to toxins might also show comparatively sudden surges or dips in line with the rates of exposure within the population. Both childhood cancer in general, and acute leukaemia in particular, appeared to become substantially more prevalent in the years after the Second World War, and researchers attempted to deduce whether these changes were the product of improved registration and recognition, or reflections of new risks affecting the young.

The postwar growth in registered cases of cancer is most clearly presented in a 1981 history of cancer registration. The report compares figures for 1948 and 1978, and highlights two key shifts in patterns of childhood mortality. In 1948 there were 340 deaths from cancer in children under the age of ten, while almost ten times as many children in the same age bracket died from infections. Thirty years later there were 1,640 recorded deaths from cancer in children under ten, and only 500 deaths from infection. While deaths from infection fell dramatically after 1948, cancer incidence seemed to multiply fourfold over 30 years (Great Britain Advisory Committee on Cancer Registration, 1981, p. 11).[13]

It is not the case that cancers became massively more prevalent; while there has been some increase year on year, most of this dramatic rise over 30 years was the result of better recording, with an increasing proportion of the country reporting cancer cases in all ages and more hospitals correctly diagnosing leukaemia or tumour (Cancer Research UK, 2012). Extra instances of childhood cancer were detected, as the condition became an expected illness, one actively looked for, a subject of scientific study.

Local registries of patients treated for cancer at the Radium Institutes operating in major British cities had been maintained since 1929, to enable the health records of people treated with the new technology of radium to be followed. In the late 1930s, as Parliament debated a Cancer Act, the aims of cancer registration broadened: it was planned to keep records of every patient treated, to investigate how the disease progressed with, and without, the treatments then available. The Cancer Act of 1939 was never implemented, but aspirations for cancer registration became more detailed over the war years. A nationwide scheme, established in 1945 by the Radium Commission and taken over by the General Register Office two years later, was designed to collect data to enable the planning and evaluation of services for people with cancer. It was hoped the register would elucidate the incidence and distributions of different cancers, providing clues as to their causes.

By comparison, in the USA registries for patients with tumours were set up on a state-by-state basis, beginning with Connecticut in 1936. However, funding for their maintenance was minimal, and few states managed to create thorough records until the 1950s, when large amounts of federal money were made available for cancer research (Patterson, 1987, Ch. 5; Aronowitz, 2007, p. 64). A national register spanning the whole of the USA was not instituted until 1973, and did not receive systematic and consistent reports from individual hospitals and states until the 1990s; those interested in rates of cancer in children in the USA had to build their own data sets.

In addition to the British national register for all patients suffering from cancer, there were also specialised registries for cases in the paediatric population. In Chapter 1 we discussed the benefits that flowed from concentrating cases of childhood cancer in a small number of hospitals, fostering the formation of highly knowledgeable teams. The Manchester group of interested paediatricians and radiotherapists was well supported at a senior level within the medical school and the various therapy departments. It was this convergence of accumulated knowledge and interest that led to the creation of the first population-based paediatric cancer register.

While working at Pendlebury in the late 1940s, Henry Basil Marsden decided to leave paediatric clinical work behind and move into paediatric pathology. Marsden had the support of his boss, Professor Wilfrid Gaisford – the leading paediatrician discussed in Chapter 1 – who was involved in MRC working groups. In 1951, Gaisford arranged for him to spend a year working as a pathologist with Sidney Farber in the Boston Children's Hospital, an experience that inspired Marsden to focus his future work on tumours found in children. Following his return in 1952, Marsden suggested founding a population-based register of all childhood cancers across the north-west region of England, the area from which such cases might be referred to Manchester. By collecting microscopical slides and patient histories from every pathologist working in the region, Gaisford hoped to build a complete picture of incidence and clinical data that would allow the varying treatments offered in different institutions to be evaluated. Gaisford, together with the Professor of Pathology, Alexander Colin Patton Campbell, and Edith Paterson, secured a small grant from the Victoria University of Manchester Student Rag charitable fund and established the tumour registry in 1953. Thereafter, the university made small occasional grants allowing this work to continue.

The register had two important side-effects. The sending of slides and samples to the pathology department at Pendlebury increased the referral rate of cases to that hospital and thereby to the Christie, building up their regional service. Also, because theirs was the world's first population-based record of cancer incidence and outcome in children, the Manchester pathologists, paediatricians, and radiotherapists received visitors from numerous American paediatric cancer centres wishing to see how the register functioned and utilise its data in research. Manchester's registry was soon copied by other regions in Great Britain and by hospitals in the USA.[14] Additionally, as we shall see below, in the early 1950s a group in Oxford began to study the incidence of leukaemia, extending their interest to all childhood cancers later the same decade.

Cancer researchers and epidemiologists noted a postwar increase in the incidence of cancer in children within a few years of the establishment of the NHS and of the national and local registries. Part of the increase could be explained away as being the consequence of more cases reaching the attention of the medical profession. From 1948, the advent of the NHS provided free treatment at the point of delivery; prior to 1948, many children with cancer may not have seen a doctor or nurse until they were dying from the complications of their malignancy, which may have gone undetected even at death. In addition, the likelihood of an affected child reaching the clinic of a cancer specialist also increased over this time period. As medical training developed to capitalise on the new arrangements for hospital care, and as paediatrics became more widely established as a specialty, postgraduate training centres in the larger cities offered ward experience in nearby children's hospitals, thus forming connections with small district hospitals in outlying areas from which difficult cases would be referred to the teaching hospital. A proportion of the increase in cases recorded in the national register, then, could be put down to being just that: a surge in the registration of childhood cancer.

Despite this explanation in terms of better registration, epidemiologists could not completely account for the rising numbers of children presenting with tumours and leukaemia. Cases of acute leukaemia, in particular, were being reported in alarming new incidence patterns. The illness seemed to be genuinely becoming more common, especially in the very young, suggesting the presence of a novel health threat.

Fear of the Bomb

In the early 1950s, the accumulation of evidence that radiation could cause leukaemia in people was prompting increasing concern within the British and American governments. By 1955, the MRC had received scientific reports demonstrating that civilians in Japan who had received nonfatal doses of radiation from nuclear explosions, hospital patients who had been given courses of radiotherapy, and communities living in areas with high concentrations of radon gas, were all found to show rates of leukaemia incidence in excess of that in the wider population. At a time when nuclear weapons were being routinely tested above ground and new nuclear energy-generating facilities were being developed, there were good grounds to fear that elevated radiation levels might precipitate a leukaemia epidemic.

On the morning of 6 August 1945, the world's first atomic bomb was detonated in the air above the city of Hiroshima.[15] Three days later, a second bomb exploded above an industrial area on the outskirts of Nagasaki. Shortly afterwards, news began to reach the USA about the impact of these bombs, not just the immediate or short-term deaths, but the health effects on survivors. American military commanders sent teams of researchers to Japan to investigate.

It was soon realised that research into the health effects of ionising radiation would be better conducted by a body not associated with the military powers who had forced the surrender of Japan. Those who had survived the bombs but suffered ill-effects, such as nausea, hair loss, and skin burns, were reluctant to cooperate with army scientists affiliated with the occupying power. For this reason, the National Research Council (NRC) of the American National Academy of Science inaugurated a Committee on Atomic Casualties, and asked the President of the USA to authorise the foundation of an Atomic Bomb Casualty Commission (ABCC). The NRC would direct the Commission's research and bring together all earlier research projects conducted by the American army and the Japanese authorities. It would also initiate larger combined studies, in order to fully assess the long-term genetic, developmental, and haematological effects of the two atomic bombs.[16]

The Commission started small but expanded quickly, hiring Japanese doctors to work alongside young physicians and statisticians from the

USA. Thousands who had lived in Hiroshima and Nagasaki were called for examination, at which blood samples were collected, bodies were measured, and full histories of family illness were taken. Participants were asked where they were at the exact time of the bomb, the degree to which they were shielded from its effects by buildings, and the health effects, if any, they had suffered in the subsequent weeks. Children, including the unborn developing *in utero* at the time of detonation, and women who subsequently became pregnant, were of particular interest to the Commission, offering a chance to investigate the long-term consequences of exposure to bodily systems peculiarly vulnerable to radiation's damaging effects by virtue of their potential for rapid cell division. By 1951, each clinic was seeing 10–20 people a day, calling many back year after year to build up a long-term picture of the health effects of those two fateful mornings.

The Commission also collected records of all the cases of leukaemia notified to the Japanese health authorities, among both the exposed and unexposed populations, throughout the country. After the first five years it became clear that bomb survivors showed an excess of cases of most forms of leukaemia, across all age groups, and especially among those closer to the detonation sites. By 1956, records suggested that the risk of contracting leukaemia was raised by a factor of at least four, and that this was almost certainly an underestimate, as many thousands of the most heavily irradiated had not lived long enough for their marrow to manifest malignant changes.[17]

Distressingly, children appeared to be even more sensitive to this fatal form of radiation damage than adults. Details of suspected deaths from leukaemia were slow to be confirmed by the ABCC, as many parents were reluctant to allow autopsies and some Japanese doctors became deeply suspicious of American motives once Japan was no longer under occupation. Some cases of childhood leukaemia also appear to have initially been classified as pernicious anaemia until tissue samples were later re-examined for histological diagnosis. Nonetheless, ABCC statisticians discovered the leukaemogenic effect of radiation was far greater in children exposed to the bomb than in other age groups, the risk of contracting leukaemia increasing 26-fold for children under ten who were within 1,500 meters of the bomb (Miller, 1968).[18]

In 1955, as the first ABCC results demonstrating that high doses of ionising radiation had this marked leukaemogenic effect in both adults and children were published, the MRC was also facing evidence of a significant increase in the incidence of leukaemia in Great Britain (FD 1/8100).[19] Oxford University boasted the country's most highly regarded group researching disease patterns in the population, the Social Medicine Unit. David Hewitt from this Unit prepared a memorandum on leukaemia for the MRC, suggesting that an apparent increase since the late 1920s could not be explained purely by better diagnosis or improved reporting. The rise in incidence was not seen in every country and was therefore likely to have an environmental cause; a report of this nature from the respected Oxford group could not be dismissed.

The Social Medicine Unit had been established, originally as the Institute of Social Medicine, in 1943 under the leadership of Professor John Ryle, who left the prestigious Regius professorship at the University of Cambridge to accept the role.[20] A leader of the social medicine movement that had considerable influence in the 1940s, he strongly supported the foundation and principles of the NHS. While many social medicine experts were concerned with public health and health policy, Ryle's unit turned towards clinical medicine and gained a reputation for pursuing aetiological investigations into medical conditions that were exciting the most clinical interest (Lewis, 1992).

Researchers in the Social Medicine Unit were growing increasingly concerned about the effects of ionising radiation, present naturally in the world but also increasing through medical uses of radiation and the atomic weapons programmes of the USA and other countries. Hewitt reported that a new pattern of leukaemia incidence by age had appeared circa 1950: markedly more children between the ages of two and five were dying from leukaemia than had ever been seen before; indeed, these children seemed more prone to developing the illness than any other age group bar the most elderly.[21] There were fears that the Oxford group had detected the first wave of a new epidemic, perhaps triggered by some recently introduced leukaemogenic agent.

The same rise was evident in Scandinavian countries, as Hewitt noted, and also in the USA. Reports of the steep curves in incidence rates over time appeared in American newspapers in the late 1940s, along with speculation that this was purely a response to the expansion in the number of doctors specialising in detecting and treating blood diseases over the same time period (review *Blood* in 1947 and news article entitled 'The Leukemic Terror' in *Newsweek* in 1949, discussed in Wailoo (1997, p. 168)).[22] Hewitt's memo demonstrated that, at least in the case of Great Britain, no such explanation could be used to flatten the incidence curve. The increase was real.

In Edinburgh, radiologist William Court Brown was investigating claims that radiation administered to sufferers of ankylosing spondylitis, a form of inflammatory arthritis, had led to some developing leukaemia; the immediate stimulus was a claim from a widow for compensation from the Ministry of Health. Court Brown had moved from clinical practice to full-time research in the 1950s, joining the scientific staff of the MRC and working in state-of-the-art laboratories built to support his group's studies on the clinical effects of radiation.

At the same time as Court Brown was writing up his discoveries, epidemiologist Eileen Woods was submitting a report to a Panel of the MRC's Committee on Protection Against Ionising Radiation on an excess of leukaemia deaths in Cornwall, apparently correlated to the high levels of radon encountered in the region: those who lived in homes built above seams of granite appeared to be excessively prone to developing the bone marrow cancer. In response to these four sources of troubling information, the MRC called a conference on leukaemia, to be held in London on 17 December 1954.

Through the autumn of that year, Professor Leslie John Witts, an expert haematologist employed in the world-leading haematology department at the Radcliffe Infirmary in Oxford, debated by letter with Sir Harold Himsworth, Secretary to the MRC, about who should be invited (FD 1/7823).[23] Witts was opposed to calling a meeting of the full MRC Haematology Panel; he wanted to bring together a smaller group of people, a mix of the leading researchers on radioactivity and experts in toxicology, who together could advise the country's leading clinicians specialising in blood disorders. Finding support from Frank Green (Principal Medical Officer at the MRC's London headquarters), Witts proposed the group not only discuss the radiation studies, but also explore the possibility that Hewitt's findings could be explained by the introduction and increase of synthetic chemicals both in industrial use and food.

Eventually, invites were sent to 20 people, and copies of the papers by Hewitt, Woods, and Court Brown were precirculated. At the conference, participants agreed on several recommendations for the future direction of work on leukaemia. Cases notified to the Register General's office were to be specified as acute or chronic, and, where possible, as myeloid, lymphocytic, or monocytic.[24] Although agreeing with Hewitt that infection might precipitate the onset of leukaemia, the committee accepted as a working hypothesis that the most likely root cause of the increase in prevalence was irradiation.

Out of this conference, a working party was set up, charged with defining the condition and its subtypes for registration purposes, and with supervising and coordinating research into leukaemia's aetiology. The element of Hewitt's report that was most troubling was the new peak in cases seen in children between the ages of two and five (Hewitt, 1955). The minutes of the conference record the first plans for what became the Oxford Childhood Cancer Survey:

It was thought that a careful enquiry involving the parents of children with acute leukaemias (and controls) might possibly give new information on aetiology, just as the original Hill and Doll investigation in regard to lung cancer (FD 1/7823).[25]

In 1950, Austin Bradford Hill and Richard Doll, working in the MRC's Statistical Research Unit, had undertaken a preliminary study of possible reasons for the apparent increase in lung cancer. This had quickly revealed a connection with smoking, and had been followed by a case-controlled study, comparing people with lung cancer with matched controls, one control selected for each cancer patient who as closely as possible came from the same birth cohort, the same class and region, and the same sex. This study strongly suggested a causal connection between smoking and lung cancer: not merely a correlation, as had been argued by defenders of the tobacco industry.[26] Case-controlled studies had thus proved themselves capable of revealing causal connections between suspected carcinogens and the onset of cancer.[27]

Acute leukaemia in children was especially promising as an area of research into the cause of the cancer because the patients had short and relatively simple histories. There was little chance they had suffered exposure to toxic substances used in some workplaces, and children suffering from the disease in the first decade of life had experienced a relatively small number of prior infections. The findings of the ABCC were suggesting that leukaemia followed exposure to ionising radiation after a latency period of 4–7 years. The ABCC was itself monitoring the development of children born to irradiated parents to see if a genetic effect passed from parent to child. The possibility of genetic effects from *any* pollutant could be explored by studying not just the medical history of leukaemic children, but also that of their parents, preconception. Thus, the working party commissioned a study of leukaemic children designed to identify possible causes, not just in infancy, but also in antenatal maternal exposure to carcinogens.

The Oxford Childhood Cancer Survey

Alice Stewart was one of a small number of practising epidemiologists in the early 1950s, having moved from clinical work into epidemiological research in 1945 when appointed assistant to Ryle at the Oxford Institute of Social Medicine. Stewart became head of the group on Ryle's death in 1950, but the Institute was downgraded to a Unit and Stewart was therefore not promoted to professor; as Stewart was the daughter of Professor Albert Naish, the first holder of the chair in paediatrics at Sheffield, this slight may have caused considerable disappointment.[28] Like her colleague Hewitt, Stewart was curious about the increase in leukaemia.

Witts, Green, and Doll were all keen to get some aetiological research underway immediately in order to 'have something for the working party to bite on' (FD 1/7823).[29] Doll and Witts visited Hewitt and Stewart in February 1955 and were optimistic the Oxford Unit would perform the investigation capably. The working party itself, including Court Brown, met in March to review his study on ankylosing spondylitis patients and discuss the design of the investigation: how families would be sampled, what questions they would be asked, and how much the work would cost (FD 1/7824).[30]

Stewart began a case-controlled study of all recorded deaths from leukaemia and other malignant disease in children up to the age of ten. Taking mortality records from 1953 as her starting point, she visited all the 203 medical officers around the country herself, persuading them to report to her each leukaemia death in children under ten there had been in the area. Families were contacted and interviewed. Controls were located for each case and also interviewed.[31] She was subsequently criticised for allowing each region to use its own interviewer to collect the mothers' reports, but argued to the working party that this approach was necessary when collecting data over so wide an area.

Stewart presented her preliminary findings at the second meeting of the working party in June 1956; although the number of returns that had come in was small, there was a strong suggestion that leukaemia was more common in children whose mothers had had X-rays to their abdomen during pregnancy. The group agreed that if further reports substantiated this correlation – there was no certainty about any causal connection at this date – then a note in the general medical press should be considered, alerting obstetricians to the possibility that there might be a cause and effect relation (FD 1/7826).[32] Stewart's work appeared to confirm Court Brown's findings on patients given X-ray therapy for nonmalignant disease, but was even more disturbing: not only were the doses used in obstetric X-rays much smaller than those given to sufferers of ankylosing spondylitis, the practice was comparatively common. In what was the peak year for above-ground nuclear testing and radioactive fallout, Stewart's results – that a single dose of X-rays in early pregnancy more than doubled a child's risk of developing leukaemia – flew in the face of determined reassurance about the safety of low doses of radiation coming from obstetricians, and from the advocates of nuclear weapons (Davis, 2007, p. 415).

The leukaemogenic potential of radiation had been under investigation since the 1920s when radiologists, and workers with radium, were first noted to be unusually susceptible to cancers in the bone marrow and the bones themselves. But in 1956, evidence was mounting that the risk of contracting cancer was not limited to those who worked with radioactive materials; it could also be iatrogenic, a result of radiation used in medical treatment.

Concern over the hazards of radiation was not confined to medical researchers and practitioners. The general public was agitating for more and better information in the face of alarming facts and suggestions published in the media about the consequences of working with radionuclides, of radiotherapy, and of fall-out from the atmospheric testing of nuclear devices. On 29 March 1955, the Prime Minister, through Lord Salisbury (the Lord President of the MRC), requested the MRC appoint a committee to report on the hazards of radiation. The committee, which included Himsworth and Bradford Hill, brought together men in the most senior positions in radiology departments within teaching hospitals, men who led MRC research groups exploring radiation's effects on animals, and the Council's usual experts on epidemiological method. Court Brown was appointed to the Secretariat.

The committee's report, *The Hazards to Man of Nuclear and Allied Radiations*, was duly published in 1956. Doll and Court Brown produced an appendix on the incidence of leukaemia among survivors of the atomic bomb explosions:

the observed incidence among survivors who were less than 1,000 metres from the hypocentre is 100 times greater than the mortality which would have been expected (Medical Research Council, 1956, p. 85).[33]

Doll also expressed concern over the continuing programme of nuclear device atmospheric testing. In a paper he prepared for the Individual Effects Panel of the MRC's Committee on the Medical Aspects of Nuclear Radiation in 1955, Doll estimated that fallout from the existing nuclear test programme would result in an additional 900 deaths each year in the USA, and an extra 100 in Great Britain (FD 1/8100).[34]

The foundation of a research programme into the aetiology of leukaemia in the mid-1950s, therefore, can be seen as the MRC's strategy for estimating the likely scale of further increases in the rate of leukaemia. But knowing how many cases there might be and what had caused them was not enough in itself. Knowledge of causality had to be supplemented with efforts at prevention, and better therapeutic options for those who fell foul of the effects of the nuclear age. Complete prevention was not possible. The MRC working parties on leukaemia, and on the medical aspects of radiation, accepted that their guidelines for 'safe' exposure limits to radiation had to fall in line with those set by the national and international atomic energy agencies and radiological protection bodies, and that naturally occurring radiation could not be removed. The most the MRC could add to the advice already set by those governing the military and energy uses of nuclear power was to advise restricting the use of X-rays to cases where it was a medical necessity. *Hazards to Man* recommended an immediate phasing out of the use of X-rays for children's shoe fitting, along with substantial reductions in the use of radiotherapy for the treatment of nonmalignant conditions in all patients.

Attention turned to what could be done for those who contracted leukaemia. Advances being reported from American research centres, and developments within the few laboratories in Great Britain testing chemicals for antileukaemic potential, suggested that new agents might afford sufferers longer survival times. Witts and Himsworth agreed the existing working party on the causes of leukaemia could be left to 'look after itself' as it had 'such an impetus' to investigate the role of ionising radiation in the rise of leukaemia (FD 1/1337).[35] A second conference was called in the summer of 1957 to consider instigating trials to clinically test the new drugs. However, before trials could be considered the skills of pathologists and clinicians in distinguishing the various types of leukaemia needed to improve. Himsworth strongly supported the notion of forming a working party purely to make 'an organised onslaught on the typing of leukaemia', as without this 'our studies on causative factors and therapeutic trials must inevitably lack precision' (ibid). A 1955 World Health Organization report had revealed huge differences between countries in the proportions of leukaemias classified as acute and chronic, and as lymphocytic and myeloid, indicating that an inability to discriminate in clinical practice was widespread. Thus, in the latter half of 1957, two new working parties were formed, one to tackle the thorny issue of distinguishing leukaemia types, the other to begin planning clinical trials of the chemotherapeutic agents emerging from American and British research. We will pick up on the work of these groups again in Chapter 3.

Antibiotics and the Visibility of Leukaemia

There is a close relationship between the introduction of antibiotics to children's hospitals and the emergence of childhood cancer, and leukaemia in particular, as a major threat to health. We have already seen that antibiotics led to a larger slice of the mortality pie chart falling to cancer. But the relationship went deeper. Stewart's survey indicated that the increase in incidence of acute leukaemia in children and the new peak in incidence between the ages of two and five were not due to any new toxic agent in the environment. Instead, they were the result of children being diagnosed with overt acute leukaemia who, without antibiotics, would have died from infection before leukaemia had been diagnosed. As Stewart, with her colleague George Kneale, described in 1970:

> more than half of the cancers which present before the age of ten are haemopoietic neoplasms whose effects on immunological mechanisms are so devastating, and expressed so early, that it is only in privileged communities and in recent times that their true incidence has been revealed (Stewart and Kneale, 1970a).

This simple conclusion was drawn from painstaking analyses of the Oxford Survey's bank of children's medical histories, collected from the mothers of over 7,000 children under the age of five between 1955 and 1962. Half of these children had died from leukaemia or a solid tumour, while the other half were healthy, included in the survey as case controls in order to help epidemiologists identify significant differences in medical histories between the two groups. Kneale (1971) identified that leukaemic children were about twice as susceptible to serious attacks of infectious diseases in the year before they were diagnosed, such as pneumonia and bronchitis, than the healthy control children. Using national cause of death records, Stewart and Kneale (1969) plotted two graphs of death rates over time, the first for pneumonia, sloping downwards from 1.5 per 1,000 people per year in the 1910s to almost zero by the end of the 1950s, and the second for leukaemia, sloping upwards from around 1 per 100,000 people per year over the same period. This strong correlation supported the theory that developing leukaemia left children less able to fight off infections, and that the advent of antibiotics had enabled more sufferers to live long enough for their underlying condition to be recognised.

Children diagnosed with leukaemia today have often been taken to a family doctor or hospital accident and emergency department following a few weeks of pallor and tiredness, or bruising and blood blisters, or persistent and repeated infections, or some combination of these symptoms. Each symptom is caused by a shortage in the child's body of the relevant kind of blood component. Before antibiotics were available, many children would have succumbed to an infection. Unable to fight it owing to low levels of neutrophils, their deaths would be recorded as caused by infection, the underlying leukaemic condition escaping detection.[36]

Stewart's initial question had been: What had caused the increase in incidence in leukaemia in children? The link she discovered between exposure to X-rays *in utero* and developing cancer in the first decade of life accounted for an additional 40 cases of childhood cancer out of the 1,500 investigated, about six per cent of the total.[37] Her quest for an environmental cause ultimately led to a different finding: the peak in susceptibility between ages two and five had been there all along, masked by infections.

Treatment for Leukaemia Sufferers in the 1950s

Little could be done for children with leukaemia in the early 1950s. Blood transfusions were occasionally given in cases of extreme anaemia, but should a child's level of platelets decrease to the degree likely to produce haemorrhage, there was no remedy. Antibiotics might keep opportunistic infections at bay, but the disease process could not be halted. Death followed within a few months, sometimes more quickly, as many children were not diagnosed as suffering from anything more than lethargy and infection until leukaemic cells had infiltrated their organs.

However, in Boston, New York, Washington, and Buffalo clinicians and scientists were collaborating in the development and clinical testing of drugs that showed promise against leukaemia. Steroids, antimetabolites, and alkylating agents appeared capable of temporarily arresting the disease process, buying the child a few months of good health, albeit in return for unpleasant and dangerous side-effects. Some British clinicians, if they could secure supplies, were trying the same drugs in isolated adult or child patients, and by 1957 the MRC working party was considering scaling up research into controlling leukaemia.

Leukaemia had emerged as an attractive cancer for investigation when research into the effects of nutrition on anaemia was applied to this malignancy. The paediatric pathologist, Sidney Farber, chose to study the effects of metabolic agents on leukaemia after reading about advances made against pernicious anaemia in the early 1940s. In 1943, a group of researchers exploring the connections between nutrition and cancer had published papers demonstrating that folic acid could inhibit the growth of some tumours in rodents (Piller, 2001, p. 289). Having administered folic acid to his leukaemic patients, Farber found to his horror that the B-vitamin actually accelerated the disease progress. He therefore asked chemists at Lederle Pharmaceuticals, led by Dr Yellapragada SubbaRow, to synthesise a chemical blocker for the vitamin, something to be drawn into the cancer cells and make folic acid unavailable to them, thus decreasing their ability to replicate. Farber immediately applied the chemical to children with metastatic cancer and in 1948 stunned both chemotherapy researchers and experts in haematology when he published his findings that aminopterin, a chemical synthesised at his request, could bring about a temporary remission in children with acute leukaemia (Farber et al., 1948).[38]

Farber's demonstration that the progress of acute leukaemia could be halted by the administration of a chemical compound, even if only temporarily, would not only revolutionise care for patients suffering from leukaemia, but also the treatment protocols for most cancer patients. In Chapter 3 we trace the expansion of research into chemotherapy for larger numbers of leukaemia patients utilising a wider range of chemical agents in the 1950s. In Chapter 6 we examine how similar techniques for clinically testing chemotherapy were applied to those suffering from intractable and dispersed cancers in the 1960s and 1970s, and how this contributed to the establishment of a fully professionalised specialism of paediatric oncology, covering the complete spectrum of childhood cancers, by the end of the 1970s.[39]

3
Working with Larger Numbers: The Development of Large-scale Clinical Trials

Cancer in the young was both increasing in incidence and becoming more visible to those in the medical professions during the 1950s. At the same time, early small-scale drug trials in the USA were demonstrating that chemotherapy might have potential in the treatment of leukaemia. Intriguing results from isolated handfuls of test subjects drew medical and political interest, and cast acute leukaemia as a disease ripe for a big push; within a decade of Farber's first successes with children in Boston, the USA could boast highly organised clinical research networks. Through these, medical teams could secure substantial grants or contract monies to acquire and maintain the equipment and facilities required for treating and assessing hundreds of patients – some of them children – and testing chemotherapeutic cocktails.

The situation in Great Britain was more constrained and contentious, with fewer resources available and more debate concerning fairness: the fairness of directing doctors to randomise their patients to different trial arms; the fairness of limiting patients' access to experimental drugs; and the fairness of asking children and parents to suffer additional months of pain and distress without long-term reward. In the late 1950s, when clinical trials of chemotherapy for childhood leukaemia were eventually launched in Great Britain, the programme met with strong resistance, even amongst elite haematologists and leading paediatricians; the trials seemed to ask too much of families and doctors for too little hope of benefit. It was reference to the American system of cooperative groups and the value of teaching new skills and better discipline to clinicians that sold the argument and got the British trials programme underway.

Postwar Changes to American Research Culture

The American government financed cancer research to a degree that was simply not possible in Great Britain, given the levels of debt the country was carrying at the close of the Second World War. As we saw in Chapter 2, Congress awards to the coffers of the National Cancer Institute (NCI) rose more than a 100-fold in the late 1940s, from $180,000 in 1945 to $4 million in 1947, and to $46 million in 1950. The American Cancer Society pushed the cause to Congressmen and Senators concerned with maintaining voter support. There was optimism that money would yield results: American wartime research efforts in developing new compounds effective against malaria and other infections had demonstrated that organised research programmes could quickly deliver powerful new weapons against disease (DeVita and Chu, 2008).

The creation of the Sloan Kettering Institute in 1945 within the cancer-specialist Memorial Hospital in New York serves as an illuminating case study. Generously endowed by Alfred Sloan, President of General Motors, and named by him after Charles Kettering, Director of Research at the same company, it was modelled from inception on industrial research divisions. Cornelius Rhoads, who had been Director of Research at Memorial since the beginning of the Second World War, was tasked with populating the new institute with keen scientists who could work together in the fashion envisaged: applying basic science to the control of cancer. Rhoads had worked on chemical weapons research during the war, and had already witnessed tumour regressions as a result of mustard gas application. Searching for cell-killing action, the Institute took thousands of chemicals synthesised by pharmaceutical companies for testing on specially bred tumour-bearing mice. The most promising were then tested in the wards of the hospital where terminally ill patients with little to lose might be offered the chance to serve as test subjects (Bud, 1978).

The Institute for Cancer Research in Philadelphia followed the same approach. Founded in 1933 by Samuel Fels, President of a large soap company, the Institute embarked on a programme to test already available compounds against mouse models of cancer, and was also structured as an industrial research enterprise. Three other major centres of cancer research were in operation in the postwar years, where drugs that performed well in the laboratory could easily be moved into clinical use to assess their value in treatment: the laboratories and hospitals clustered around Harvard, in Boston and Cambridge, Massachusetts; Roswell Park in Buffalo, New York, established in 1898 and the first research institute in the USA to focus solely on cancer research and treatment; and the NCI (Keating and Cambrosio, 2002).

Great Britain also had a small number of coordinated chemotherapy research groups in operation in the 1940s, with the Chester Beatty Institute (formerly the Institute for Cancer Research) in London at its core. The Institute was founded in 1909 as the research laboratory for the London Cancer Hospital, and renamed in 1939 in honour of its benefactor, the mining magnate Alfred Chester Beatty, an American millionaire who had chosen to become a naturalised British citizen in the 1930s. In 1946, pathologist Alexander Haddow (later Professor Sir Haddow) was appointed Director of the Institute, whereupon he instigated a highly successful chemotherapy research programme that led to the development of several important drugs for combating common cancers. However, the process of turning laboratory results into clinical innovations in 1940s Great Britain was complex, as funding for scientific investigation of cancer was awarded and administered separately from that earmarked by government and charitable sources for developing new treatments. Opportunities to deploy experimental drugs in hospital wards were hard to create and few in number until the end of the 1950s (Bud, 1978).[1]

Each of these American and British research centres was pursuing its own agenda and operating with its own definition of what constituted a 'response' promising enough for a drug to be tested at the next level. In the early 1950s, with its dramatically expanded brief and funding, the NCI initiated a project to bring the American centres' efforts together, by creating its own Clinical Center and cementing existing connections between researchers in a more formal way. The process prompted America's leading cancer researchers to debate explicitly and agree on experimental methodologies, understandings of cell biology, and definitions of disease response, making comparison of results possible for the first time; the agreements they reached also shaped the Medical Research Council (MRC)'s chemotherapy research in Great Britain. While the NCI's plan to bring centres together was designed to serve the fight against cancer in general, its implementation does not simply serve as background for the history at hand: the specific form of cancer chosen for special attention and support was acute leukaemia.

The American Clinical Trials Programme for Acute Leukaemia, 1953–59

In order to assess clinically those drugs that had shown promise in the laboratory, the NCI established and funded an ambitious chemotherapy programme in 1955. To manage this required both a Cancer Chemotherapy National Service Center, to oversee the testing of promising chemicals in laboratories across the country, and a Clinical Trials Cooperative Group Program, to support the creation of a regionally organised network of groups through which clinical researchers could pool their efforts in testing the best compounds in cancer patients.

While cooperative groups are now deemed essential when dealing with rare conditions in order to recruit significant numbers of test patients, as Robert Bud (1978), Ilana Löwy (1996), and Peter Keating and Alberto Cambrosio (2002) have shown, this is a managed necessity, not a foregone conclusion. Mid-twentieth-century models for scientific medical research suggested investigators should initially seek to understand the fundamental cellular mechanisms involved, and then design or identify treatments to address a well-understood disease process. However, in cancer chemotherapy trials, scientific investigation and empirical testing of treatments ran side by side, the two mutually informing one another's direction, in order to advance knowledge and improve patient outcomes simultaneously. In its early years, cancer chemotherapeutic research was heavily criticised by heavyweight clinical teachers, precisely for being empirical, for proceeding before the scientific bases had been properly understood. However, those engaged in the endeavour were quick to point to favourable parallels with other diseases controlled through chemotherapy, such as tuberculosis and pneumonia: what mattered was that a cure was found; one could take time to understand and explain its mode of action later.[2]

Cooperative groups brought researchers and clinicians in Boston, New York, Houston, and elsewhere into a coordinated search for sequences and, later, combinations of drugs that would bring cancer under control. Working in concert, using the same means of selecting and allocating patients to particular arms of an agreed and shared treatment protocol, groups of clinicians exchanged their individual expertise and judgement to generate collectively results that others, in government and in distant hospitals, would see as objective evidence for or against new cancer treatments, and to secure continuing funding for further research (Marks, 1998).

By 1959, there were seven main groups conducting trials of chemotherapy for leukaemia or solid tumours, but the patient base consisted mainly of adults. The first two study groups taking children as patients were set up in 1955, and labelled Acute Leukemia A and Acute Leukemia B, which, in 1964, changed its name to Children's Cancer A.

Other groups were established throughout the late 1950s, defining themselves either by the cancer they were working on or the geographical region from which participants were drawn. The Southwestern was one of five groups to start work in 1956 and, apart from Acute Leukaemia B, was the only group to be paediatrically focused before the late 1960s.[3] In 1959, acute leukaemia was chosen as the one cancer on which efforts should concentrate owing to the ease with which effects of various candidate drug regimes could be measured: levels of leukaemic cells – in patients' blood supply and bone marrow – were reliable indicators of disease level, and leukaemia in humans appeared to be well-modelled by leukaemia in mice, suggesting that compounds that prolonged life in leukaemic mice had a reasonable chance of behaving similarly in human sufferers. As there were no efficacious therapies available for acute leukaemia it was also deemed an ethically suitable choice; there would be few moral problems in offering treatment that may or may not slow the disease process to patients for whom nothing else could be done. Testing chemotherapy as a first treatment has proved harder to justify for cancers where alternatives have existed: withholding surgery, radiotherapy, or hormonal therapy from patients who would likely benefit, and offering them alternative treatment of unknown value, has seemed unwarranted to many clinicians; it is hard to posit such experiments as being in the patient's interest (Timmermann, 2008). Testing chemotherapy after the use of other treatments makes interpretation of results harder, as investigators have to distinguish between the effects of the various therapeutic modalities deployed when assessing any good response observed.

By the end of the 1950s, then, scores of American clinicians and hospitals were working together in these networks, employing the same protocols – tables and flow charts that stated which drugs should be given at what doses and when – and randomising their patients between the various experimental arms, in order to collect information on the effects of steroids, methotrexate, mercaptopurine, and numerous other chemicals that showed favourable results in animal models of human leukaemia. Many of those involved in this programme were new to the field of cancer research, yet it is clear from their memoirs that the work was gripping, leading individuals to commit extraordinary levels of dedication to changing the outcomes for cancer sufferers.[4] With a steady supply of funding for research aimed at extending survival times or producing cures, clinicians were able to create and test new protocols at a time when the British government had a rather different set of priorities for allocating monies to health services.

The Southwestern Group: Texas and Hiroshima

Hindsight can appear to reveal an impressive orderliness to the development of the cooperative groups and their first clinical trials. However, memoirs and personal communications from those involved reveal the frustration and exhaustion that setting up a clinical research unit entailed. The NCI chemotherapy programme facilitated the securing of funding to cover staff time, access to medical facilities, and supplies of drugs and other consumables, but negotiating control over beds for patients needing inpatient care, and timely reliable responses from pathology laboratories serving many different physicians, could prove stubbornly difficult. Participants' recollections of the early years of the chemotherapy trials programme reiterate that finding managerial staff willing to support the introduction of innovations, and families who could stomach submitting their children to regimes with terrifying side-effects, was not an easy process.

Historians have studied closely the foundation of the Acute Leukemia Group B. This group was centred on the team of leukaemia researchers working out of the NCI's own Clinical Center, but even here there were battles over the legitimacy of treatments proposed by the predominantly young researchers. Tom Frei and Emil J. Freireich have been hailed as heroes for their development of supportive care for patients in mortal danger from side-effects of the intensive chemotherapy under test, and for their establishment of criteria for evaluating the effectiveness of cytotoxic drugs that other groups could replicate.[5] But this was only one of seven groups up and running by 1959, and while Acute Leukemia Group B made impressive and radical breakthroughs, the largest of the groups, the Southwest Cancer Chemotherapy Study Group (SWCCSG), was making waves by researching how to effectively run cooperative trials.

The SWCCSG was founded by the paediatrics department at the M. D. Anderson Hospital in Houston, Texas, to research the potential of chemotherapy in the treatment of childhood cancers. The group was established shortly after the opening of a new department by the three paediatricians recruited to lead the unit, clinicians and skilled research administrators who had worked together in Japan. A distinctive feature of the Houston unit was their research into the best ways of conducting research programmes themselves, knitting epidemiological studies, animal experiments, and clinical trials together in a concerted assault on cancer. This group therefore deserves closer scrutiny for its contribution to the methodology of investigating potential chemotherapies for childhood cancer. The SWCCSG also commands attention for its rapid emergence: it seemed, to contemporaries, to have sprung from nowhere.

The M. D. Anderson Hospital and Tumor Institute had been founded in 1941, to treat cancer sufferers living in Texas. An initial appropriation of $500,000 from State taxes was authorised to establish a hospital 'devoted to the diagnosis, study, prevention, and treatment of neoplastic and allied diseases', funds matched the following year through the M. D. Anderson Foundation, a charitable body founded by local businessman, Monroe D. Anderson. The hospital started on a small scale, with four researchers and patients only seen in clinics, inpatient care beginning three years later using beds leased from other hospitals in the city.[6]

In 1946, the original director of the hospital, Dr E. W. Bertner, moved to direct the Texas Medical Center, the overarching management structure coordinating the work of the two medical schools and the many separate hospitals all collected together in one region of Houston. Surgeon Lee Clark took over for a long and highly successful career managing the hospital. A year after his appointment, a firm of architects was commissioned to begin work on a new and much larger hospital building, eventually opened in February 1954. In pride of place lay a cobalt-60 irradiator, built at the Oak Ridge National Laboratory, for supervoltage treatment: cutting-edge radiotherapy. A paediatric ward was scheduled to open the following year, in a space large enough to accommodate 38 beds in a set of rooms that could be custom designed and decorated for the needs of seriously ill children and their families (M. D. Anderson Hospital and Tumor Institute, 1965).

In 1953, at the time Clark was planning this new paediatrics department, the Board of Regents of the University of Texas was seeking a new Dean for the Postgraduate School. In the spring of 1953, the hospital and medical school jointly agreed to invite the paediatrician Dr Grant Taylor to take up a senior position in the institution on his return to the USA following a tour of duty as Field Director of the Atomic Bomb Casualty Commission (ABCC) in Japan. Why did a cancer hospital seek out this man with military and directorial experience? And why did Taylor accept such a post – what did he see in his past experiences as relevant to the new role, and what did he hope to gain from taking on such a large responsibility?

Prior to his wartime service in Japan, Taylor had been the Associate Dean of the prestigious Duke University Medical School in Durham, North Carolina, where he had completed his medical training and residency under the highly esteemed paediatrician, Wilburt Cornell Davison.[7] While in Japan, Taylor and his colleagues read about the new facilities planned for Texas in an issue of *Time* magazine. Taylor and two of his closest colleagues discussed the possibility of 'future professional affiliation with the Anderson', and Taylor volunteered to investigate.[8] His expressions of interest in the opportunities afforded by the rapidly expanding hospital and associated medical school soon reached the desk of Clark, who saw in Taylor a potential head of department. Clark offered Taylor a position that combined academic status, clinical power, and research freedom with a generous salary.[9]

As a paediatrician and a manager, Taylor was a strong candidate to assemble a new department and launch an innovative research programme. However, he had very little experience of dealing with cancer patients. Taylor's qualifications as an educator were indisputable; but in what way was he qualified for this new post, as head of not only a ward for child cancer patients, where previously no such provision had existed within the State, but also a new 'division of experimental paediatrics'? Taylor's reminiscences suggest he had more experience of childhood cancer than his prior medical responsibilities might have suggested: he recalled that at the time of seeing the article in *Time*, he and his colleagues were facing clear data that the ionising radiation from the atomic bombs was carcinogenic; and files from Taylor's time in Japan indicate the Commission doctors were being asked to confirm diagnoses made in Japanese hospitals, not only of the many leukaemia cases – cases that were appearing more frequently than doctors would expect given the size of the population – but also of many other childhood neoplasms.

It was not Taylor's cancer experience that he himself believed relevant to the new post, however. He later recalled that the specific skills and pertinent background he brought to Houston were his experience of and commitment to 'horizontal integration' of the many diverse research programmes run by a large research body. In the ABCC he had ensured that all heads of department were aware of one another's work, thereby facilitating cooperation; he envisaged setting up similar patterns of professional networking within the Texas Medical Center.[10] He was also passionately committed to involving families in their children's care within the hospital, and foresaw that encouraging this could make their participation in research programmes more bearable.[11]

Once in place as senior paediatrician and Dean of the Postgraduate School, Taylor moved quickly to recruit two colleagues from the paediatrics department of the ABCC, Wataru Sutow and Margaret Sullivan, Sutow being first to join him in Texas, and Sullivan arriving a year later. Over the course of his career at the M. D. Anderson Hospital, Taylor was to find temporary or permanent positions for no fewer than 132 Japanese medical people who had served in the ABCC, in addition to a much smaller number of American physicians who had also worked there.[12] Taylor was keen to keep his former team together because he had seen each of them master 'biomathematical programmatic medical research'.[13] It was this expertise that was so valuable to him and, he foresaw, to the M. D. Anderson and the NCI: these paediatricians were known to be able to conduct large-scale research programmes, designed to deliver analysable data capable of yielding significant results.

Taylor and those he worked with in Houston and the SWCCSG were not alone in stressing the need to plan research programmes exceedingly carefully. The SWCCSG repeatedly returned to the issue of ensuring each trial only asked questions about drug effectiveness that could be answered with the patient numbers, drugs, and funds available.[14] It was clear to all working in American cancer research centres that successful clinical trials programmes required discipline in trial design in order that patient responses could be accumulated across treatment sites to produce better protocols and explanations for their effectiveness.

Scaling Up: Importing the Experience of the ABCC

Wataru Sutow had joined the ABCC as head of the paediatrics programme in 1948 and, apart from one short period of civilian life working in Stanford, spent most of the next five years coordinating its work. On arrival, Sutow's first task was to contact the 300 exposed children in Hiroshima whose growth and development had initially been studied by Professor and Mrs W. W. Greulich in 1947.[15] Sutow aimed to expand the survey of the effects of radiation on growth and development to several thousand children from both bombed cities, with appropriate children selected as controls. By 1953 the paediatrics department was seeing more than 100 children each month, following up with annual check-ups on their birthdays, recording their physical details on punch cards, and publishing six-monthly reports of the data.

The struggle to gather useful data was not unique to the paediatrics department. A records committee, founded in February 1949 to design sheets on which to record data, had been charged with surveying the various record forms employed by the ABCC:

> The committee should proceed with the idea of promoting as much standardization as is practicable, avoid needless duplication, and to assure that the data collected is in its most useful form.[16]

The committee had redesigned the forms on which Japanese interviewers had recorded medical and nonmedical histories of children chosen for the Commission's studies. Forms for recording growth and development were produced on the same template: each possible answer to a question was allocated a number corresponding to a hole on a punch card. Operatives in the central offices made holes to represent all the 'yes/present' answers, and the cards were then stored compactly in metal filing cabinets, ready for automated analysis by the expensive IBM machines used for the purpose, first in Ann Arbor, Michigan, and then, from 1950, in the Hiroshima offices of the Commission.

In 1952, Sutow's chief assistant, John Wood, had prepared a report for the Commission's controlling body in Washington, the National Research Council, in which he commented on the time-consuming process of data collection:

> Up until this time a large portion of the total effort of the pediatric department has been devoted to the collection of data. At the present time our data are in the process of being coded and punched on IBM cards. Within the current year analysis of clinical information will get underway.[17]

The paediatric clinic was by this point seeing approximately 3,000 children per year.[18]

Japanese journalists, reporting on the novel scientific methods being deployed by the ABCC, appreciated, and rhapsodised about, this use of punch cards and automated calculation:

> How then is this large number of report forms being consolidated? ... The recording of the data of the report forms is done by IBM electric tabulating machine (the IBM methods was [sic] introduced in the August 1948 edition of this magazine.) The card used on this machine is 186 mm in length and 83 mm in width, of very good quality paper, and of light yellow, and reddish-yellow colors. This card is placed in the key punch machine and as the fingers run over the number keys the various items on the report form are recorded as punctured holes representing various combination [sic] of figures. The card with the holes slides out smoothly to the left and the numerous data have been compiled into a simple perforated card. When cards are placed in the sorting machine they are classified automatically according to

data such as locality, sex, etc. Furthermore, the indicator will tell you the number of cards. By this method, instead of a large filing set-up, it is sufficient to have a steel cabinet to hold the small cards which have absorbed all the data. This provides economy of space, safety, and accuracy.[19]

From a few small cabinets of cards and holes came a sequence of rigorous reports demonstrating the biological late-effects of ionising radiation on those exposed *in utero* or as children. The IBM machines helped researchers produce quick analyses of large amounts of data, to reveal any high levels of correlation, such as between delayed onset of menses and proximity to the hypocentre of an explosion.

The ABCC's use of calculating machines was not unusual; it exemplified the widespread move in the 1940s towards use of and reliance on mechanical calculators and medical statistics in research.[20] This was true both for clinical trials and for epidemiological studies; the methodological skills and mathematical requirements were the same for each field of research. When analysing the possible effects of a variable applied unevenly through a large population, statistical techniques could produce levels of 'significance' and 'confidence' from the tables of figures to show whether the variable in question had, in fact, had a pronounced or definite effect. These techniques promised to make medical research more rational and scientific by delivering stronger grounds for assertions about the risks and benefits of particular compounds or conditions.[21]

The ABCC's paediatrics team prepared many reports on the health of the children of Hiroshima and Nagasaki, drawing on data collected from thousands to present statistically strong findings. Sutow acquired the status of an expert on experimental design and biomathematical models, and on his appointment to the paediatrics department at the M. D. Anderson Hospital he aspired to launch a research programme on the same scale and model.[22]

Instituting a Research Programme in Houston

Sutow took up his post as the clinician in charge of the paediatric outpatient service on 1 July 1954, with the first patients being welcomed into the ward a mere two weeks later. Taylor wanted to embark on research immediately, using animals to explore the effects of different drugs on cancer cells, and testing the most promising agents against malignant tumours in children. In December of that year, Sutow wrote to a colleague back in Japan:

> I haven't had a decent day off in the past two months. We lost our resident some time ago and I have been carrying on all of the ward work as well as the clinic work. That plus all the paper work has kept me pretty tied down. Although Taylor seems impatient at times as far as research project goes [sic], I just can't get around physically to doing it yet. Not until we get some help on the clinical side.[23]

Developing an entire cancer research programme from scratch, Taylor and Sutow were becoming exhausted.

Initial research proposals submitted to the hospital's research board outlined projects setting out exactly how to research childhood cancer. Two were submitted on 15 September 1954. In one, entitled 'Evaluation of cancer therapy in children. I. Outline of principles of methodology,' Taylor and Sutow took care to define 'clinical survey', 'clinical trial', and other terms as they were to be used in the department. In the other, entitled 'Leukemia in children. I. Basic general outline of program', the two clinicians explained how they envisaged the various projects – methodological, epidemiological, animal, and clinical – fitting together, and what the guiding 'principles of direction for [a] multiphasic coordinated approach to the problem' would be.[24]

Four months later, a third proposal was unveiled, designed 'To formulate the general principles for documentation of clinical data on children with cancer seen by the Pediatric Department of the M. D. Anderson Hospital and Tumor Institute'.[25] That Sutow and Taylor wished to specify standard methods of collecting, assessing, and combining different categories of evidence at the outset, provides a link back to their work together in Japan. Without sound method, there was every chance that data painfully collected from patients would not be capable of answering crucial questions about disease course and response. Taylor and Sutow's attention to getting the terms, methods, and forms right before taking histories or giving injections came from their experiences of research in the ABCC. Such attention ensured that Taylor and Sutow would both influence the NCI, with central committees there seeking their opinions on trial designs and requesting they chair new projects, expanding the research into chemotherapy for other childhood cancers in future years.[26]

Taylor and Sutow's projects emphasised integration between different forms of research, and collaboration between researchers across disciplines and institutions, working to address a shared question. The concepts of a distributed study group, pursuing research across hospitals, and of a multidisciplinary medical team planning treatment for individual patients, share one substantial root: both espouse the ideal of teamwork. This concept of teamwork served the Southwestern group well, helping command loyalty from invited members and support from outside agencies whose assistance was required to further research.

In 1958, Taylor was considering requesting greater state support for the services he wished to offer children treated for cancer. In a letter to a potential ally in his bid to increase state spending on children with medical needs, he wrote:

I have 32 beds for research in childhood cancer, and these beds are backed up by a basic science program in experimental pediatrics. To give you an idea of the latter: we have a mouse colony staffed largely by Fellows. What is most stimulating is the close relationship of our basic research program to our clinical research program. We now have had almost five years of experience with carefully documented material and have arrived at a point which was our real objective in coming here; i.e. to gather sufficient material to write a textbook on childhood cancer, which we have a plan to have completed by June, 1960. You may know of the excellence of the staff working with me. Each is an outstanding pediatrician who worked with me before and during the time I spent with the Atomic Bomb Casualty Commission in Japan. I mention this only to give you an idea of the background in research and the harmony which results from a group that have worked closely together for a considerable period of time on a restricted subject.[27]

For Taylor, the time in Japan taught him not only about carcinogenesis and the value of properly designing one's research question and attendant forms, but also about the importance of the team. The same rhetoric of the team – and the associated desire to be included within it – was crucial in establishing a clinical trials programme for cancer chemotherapy in Great Britain. Clinicians valued being included in MRC working parties to design, administer, and evaluate trials, and used their inclusion as grounds for dismissing fatalistic approaches to childhood cancer.

London: Initiating a Clinical Trials Programme

Little research into chemotherapeutic approaches to cancer control was conducted in Great Britain before the 1960s (Wiltshaw, 2003, p. 23–4). American approaches to the management of children with leukaemia – administering steroids, antimetabolites, and occasionally aminopterin – were applied from the 1950s by paediatric haematologists in major centres for child health research, such as the Royal Victoria Infirmary in Newcastle and Great Ormond Street in London (Lightwood et al., 1960; Thompson and Walker, 1962). Typically, these chemotherapeutic agents were administered in isolation and at low doses, according to the best estimate of the physician managing the treatment as to what level would be effective (Lilleyman, 2003, p. 42–8). In the USA, at the same time, more aggressive or combined chemotherapies were being tested. Patients were given far higher doses in attempts to kill more cancer cells with each dose and/or received multiple drugs together or in close sequence in order to combat drug resistance.[28] These tougher treatments were felt by most British clinicians to be unwarranted, sacrificing the quality if not also the length of remission for the majority of patients in order to achieve cure for a tiny minority. Indeed, when American research groups began reporting substantially prolonged lives for leukaemic children in the mid-1960s, it provoked heated debate in *The Lancet* (1965) over whether it was ethically justifiable to expose young children to such horrific side-effects for so little chance of survival. Professor Sir John Lilleyman remembered this reaction when interviewed in 2003:

> doctors were severely criticized for meddling with children who would all relapse and die eventually ... Paediatricians are a fairly conservative bunch generally, particularly in the UK, and they were fairly slow to take up the challenge to treat children with this condition, particularly when the word 'cure' hadn't come into the debate (2003, p. 44).

Given the widespread reluctance to cause child patients and their families additional pain and distress, it is not surprising that the establishment and expansion of clinical trials of chemotherapy for leukaemia took many years. How, then, did a few advocates gain the ear of the MRC in the 1950s, and initiate the process of experimentation leading to nationwide trials for acute leukaemia?

The first steps were taken not with the aim of curing leukaemia, let alone cancer in general, through chemical means, but with the attainable goal of teaching the nation's leading clinical researchers how to operate the new machinery of the clinical trial – regardless of whether the patients enrolled experienced much or any benefit. In the summer of 1957, discussions began in London over how to manage a clinical trial of chemotherapy against leukaemia. By the spring of 1960, the first two trials were recruiting patients: one trial for adults with acute leukaemia, one for children. These early trials were not 'successes' in the sense of increasing the number of survivors: none of the participants lived past the two-year mark. They were intended to teach British clinicians and hospitals how to achieve the levels of across-site cooperation necessary to run trials that, in time, might yield longer survivals.

As we noted in Chapter 2, in June 1957 the MRC called a second conference on human leukaemia to discuss the typing and treatment of leukaemia. Two new working parties were assembled, with considerable overlap of personnel: one exploring ways to increase agreement between haematologists viewing the same slide from an affected patient as to the type of leukaemia taking hold, the other debating the possibility of a series of therapeutic trials. While the former was a small group comprising clinical and academic haematologists, the latter was larger, a collection of clinicians drawn from several leading cancer centres (FD 1/7828).[29]

The Working Party on Therapeutic Trials in Leukaemia held its inaugural meeting on 16 December 1957. There was little motivation to initiate a trial at that stage: as none of the available agents was deemed 'really a "winner"',[30] there was no consensus on what the first trial should test. However, by autumn of the following year the group was planning to open a trial anyway, purely to assess the feasibility of running clinical trials in accord with those being conducted in the USA. David Galton noted in one of the working party's meetings that a 'pilot' trial 'might be helpful in determining whether the "machinery" of a trial would work' for leukaemia sufferers in British hospitals, a view supported by John Dacie's addition that even a negative result – an experimental therapy turning out to be no better than the existing management options – would be of value, as 'it would reassure about the soundness of present procedures'.[31]

The group decided to open a small trial for adult patients; children were excluded on the grounds it might prove difficult to secure the cooperation of parents and nursing staff owing to the painful bone marrow aspirations required to conform to the trial's prescriptions for measuring both the level of disease and any response.[32] The trial recruited from only a handful of the larger teaching hospitals in London, Manchester, and Edinburgh, and started treating patients in late 1959. Three treatment regimes were compared: low-dose steroids; extremely high-dose steroids; and no steroids. Each participant was also given mercaptopurine so that no patient went wholly untreated.[33]

Indeed, this trial did deliver a negative result; the adults receiving high-dose steroids – the promising regime under test – died, on average, far more quickly than those on the standard lower dose, and more rapidly even than those receiving no steroids. Witts took the counsel of his working party literally when preparing a paper for publication on the results, arguing that 'the main purpose [in instituting the trial] was to see if we could create an organisation for cooperative trials'.[34] The staff in the MRC central offices urged him to remove this statement, and emphasise there had been some hope of the high-dose steroids being more effective than the conventional dose.[35] While acknowledging the lack of positive results regarding actual treatment, Witts also suggested to MRC staff that the paper could nonetheless positively conclude 'the machinery now available could be readily adapted for the trial of more promising forms of treatment'. MRC staff passed the matter to Harold Himsworth, the Council's secretary, who proposed alternative wording, which Witts accepted: though agreeing the paper should admit the trial would not lead to any change in therapy given to patients, Himsworth stressed that the MRC committee could claim a positive benefit:

> that the method of clinical trials devised for the present investigation is reliable and could be adapted for future comparisons as new treatments become available.[36]

A separate trial for children was opened a few months after that for adults, recruiting its first cases in May 1960. Administered by Roger Hardisty, a haematologist at Great Ormond Street Hospital, the trial compared the effects of mercaptopurine given alone, against the same drug in combination with low or medium doses of steroids. Given the difficulties in securing cooperation from hospital staff, as well as consent from patients' parents, the trial recruited extremely slowly: it took over two and a half years to sign up 40 children for allocation to the trial arms. The process of allocation was far from high-tech: to ensure that children were assigned to the trial arms randomly, each time a new patient joined the trial the next brown envelope – in one of two sequences prepared in advance by Richard Doll – was opened to reveal which treatment option would be given. One set of the envelopes was kept in Great Ormond Street, from where, it was expected, the largest number of cases would be drawn; the other was kept in Doll's office within the MRC Statistical Unit in the London School of Hygiene and Tropical Medicine.[37]

Randomisation was still a new idea to clinicians within the working party, the elite of their specialty, and Doll's advice was central to ensuring all participating clinicians and hospitals allocated patients to particular treatment options on a random basis; his presence was required at every conference and meeting, and his advice on patient record sheets and on the constitution of appropriate 'controls' for the experimental arms, was always sought and heeded.[38] The need for central control over such a geographically dispersed body of clinicians was felt acutely by the MRC: for a statistical trial to be possible, the group would have to adopt uniform definitions for leukaemia, for response, and for allocation procedures; the definitions adopted were those painstakingly reached by researchers at the NCI.[39]

This trial was far more successful than the trial for adults: 50 per cent of the children survived for 40 weeks or more, compared with only 13 per cent of adults, and there was a clear improvement over the historic life expectancy of three months for children given no treatment. While all participants had died within two years, a clear and substantial prolongation of life was gained for those children given mercaptopurine and steroids in combination. These results were in agreement with findings from the American tests of similar combinations. However, at that time, American cooperative groups were testing more complex chemotherapy combinations, containing new and powerful cell-killing agents, and including radiotherapy in some cases, alongside higher levels of supportive care in the form of germ-free wards and transfusions of blood products (Laszlo, 1995; Keating and Cambrosio, 2002).

Further British trials were planned, to incorporate methotrexate into the protocol and to test the potential of a new agent, vincristine sulfate, on those of Hardisty's patients who had received chemotherapy already and thus were ineligible for the new MRC trial.[40] A separate paediatric subcommittee of the working party on therapeutic trials in leukaemia was established in 1963, under the chairmanship of Professor Gaisford of Manchester University and the Royal Children's Hospital, Pendlebury. In order to widen the intake of patients, and move through a series of trials faster, the group set about recruiting additional clinicians from as many hospitals as possible. But the results were still extremely disappointing: no more than one per cent of children treated were being cured by the efforts of this expensive research programme. Recruitment was also still a problem: as family doctors could see no benefit in referring sick children to these centres of expertise, they were not treating the vast majority of children with leukaemia. As we will discuss in Chapter 7, the experience of receiving experimental drugs with little chance of cure and the certainty of distressing effects was extremely traumatic, and few doctors would ask families to endure being part of the trials programme.

In 1964, American cooperative groups began to report good results, two years after the beginning of two trials of different intensive drug combinations, tested in adults and some children. The specified treatment protocols, designed for maximum cancer cell kill, placed patients in great danger: as their bone marrow was rendered incapable of producing blood products they required blood transfusions, platelets, antibiotics, antifungal agents, and sterile nursing environments.[41] Patients were kept in isolation, and many died of infections or blood loss, or of relapse in the brain or spinal cord many months after completing the long gruelling course of chemotherapy.

In the autumn of 1965, Witts and Hardisty visited American centres treating leukaemia in this manner, charged by the MRC with preparing a report on the feasibility, and desirability, of importing this approach (FD 23/1326). The Ministry of Health, who would be responsible for providing the funds required for well-equipped beds and high-intensity nursing, together with the MRC, arranged a conference for 15 March 1966, to discuss beginning trials of American-style chemotherapy in a number of centres to determine whether or not nationwide provision was warranted. Manchester and London were chosen for further investment, the aim being to cure leukaemia, in at least some patients, not merely deliver a series of remissions. Exploring whether or not cures could be affected on the NCI model was presented as the duty of the British government, and the ambitions of chemotherapists, with strong government backing, rapidly began to increase.

In the following chapters, we will examine how trials over the next 15 years expanded across Great Britain, first for acute leukaemia and then for all cancers.[42] In the 1960s, most children with cancer in Great Britain were seen by medical or surgical specialists according to cancer type and local availability; paediatricians, haematologists, radiotherapists, and surgeons rarely met one another or compared treatment rationales or outcomes. Only in isolated pockets, in Manchester and London as we saw in Chapters 1 and 2, were different medical specialists coming together as teams to develop coordinated and multimodal treatments for children with cancer. The American technique of forming regionally-focused cooperative groups, enabling clinicians to share data and resources in a combined search for better patient outcomes, was consciously imported and adapted to thrive in British soil, by paediatric cancer specialists seeking better support for their research and increased rates of patient recruitment to trials. This process brought together those working on leukaemia and those concerned with solid tumours (Moscucci et al., 2009). As networks of mutual support grew, and connected with the expanding group of specialists in childhood cancer in the USA, clinicians treating children with diverse cancers came to identify themselves as 'paediatric oncologists'. Members of this self-consciously new breed of doctor formed professional bodies, expanded research programmes, and, perhaps most importantly, crafted a public image of childhood cancer as curable.

4
Cancer Microbes, the Tumour Safari, and Chemical Cures

In this chapter we focus on one particular childhood cancer to spotlight how early success stories generated support for chemotherapy research. Using Burkitt's lymphoma as a case study, we examine how child survivors shaped the emerging profession of paediatric oncology and turned childhood cancer into a human interest story that could secure significant public attention.

A novel type of tumour affecting children was identified in Uganda in the 1950s. Research in the late 1950s and early 1960s confirmed it was, indeed, a disease new to medical science, and that it was peculiarly dependent on particular climatic conditions and especially sensitive to chemotherapy. Uganda became a site of intense scientific interest for cancer researchers and for journalists, centred around the person of Denis Burkitt, an Irish surgeon who had never intended to become a cancer specialist. The lymphoma, originally christened the African lymphoma but subsequently named after Burkitt, accounted for half of all childhood cancers in Uganda, yet was almost unknown in other parts of the world.[1] Burkitt announced his identification of the new cancer in 1958 and within four years had declared the cause to be an infectious agent transmitted by insects and that some 40 per cent of cases could be cured by chemical means alone (Burkitt, 1958).[2] These findings appeared to validate two major international research programmes that were then receiving government and charitable funding in the USA and in Great Britain, one seeking evidence for the possibility of cancer-causing microbes and the other testing new chemical compounds for efficacy against forms of cancer. Medical research bodies in Great Britain and the USA competed over who should have the honour of funding his research into the promising new disease.

In this chapter, we look firstly at the long history of the search for cancer germs and the reasons behind the huge growth in state funding for research into possible microbial causes of cancer that started in 1958. Secondly, we explore the webs of interest woven, by Burkitt and others, which connected a rare tumour in Africa to high-tech cancer research centres across the world, highlighting the importance of personal contacts in establishing and maintaining new linkages between centres and research areas.[3] Finally, we discuss the media interest in Burkitt's clinical research, yielding results that amplified the hope in chemotherapy's potential, stimulated by its successful application to another solid tumour just a few years earlier. Media coverage presented cured cases of the African lymphoma as cause to embrace optimism that the chemotherapy programme of the chemotherapy working groups of the National Cancer Institute (NCI) and the Medical Research Council (MRC) would deliver marked improvements for patients suffering from some of the most aggressive cancers.

Microbes: Cancer Viruses and Their Predecessors

The belief that at least some forms of cancer were caused by a contagious entity was widespread in the eighteenth and nineteenth centuries. In 1742, for example, a proposal to build a cancer hospital in Rheims prompted petitions from local residents, fearful that proximity would put them at risk of the disease from supposed airborne infection (Pinell, 2002). As staining techniques and microscopes made it easier to mark and see tiny particles in the late nineteenth century, the first specific causal agents for a number of infectious diseases were identified, and hopes of finding a cancer germ or parasite dramatically increased.

Evidence deemed to support the theory of contagious causation was presented in a variety of forms. Anecdotal examples appeared in many letters submitted to medical journals, relating tales of cancer surgeons contracting the disease from their patients, spouses from each other, maids from their mistresses, and even a dog from its master. There was also epidemiological evidence in the form of maps produced by medical geographer, Alfred Haviland (charged by the Medical Society of London in 1868 with mapping cancer mortality within the British Isles), and the local studies of D'Arcy Power, the most renowned proponent of the view that there were 'cancer houses' that would cause inhabitants to fall victim to the disease. Haviland found districts with the highest cancer mortality to be low-lying and situated on the banks of rivers prone to flooding. Power's 1899 mapping of cancer incidences in an unnamed

English village reproduced the findings of Haviland on a smaller scale, with most cancer houses located on river banks. Power therefore speculated that the instigating organism might well be related to, and living under similar conditions as, the malarial parasite. Speculation over whether cancer-causing organisms were bacterial, viral, fungal, or parasitic increased as microscopal evidence mounted from laboratories seeking the cause of cancer. Various 'bodies' were presented as the cancer parasite, often named after their discoverer. The most lauded of these cancer germ researchers were William Russell, a lecturer and pathologist at the Royal Infirmary, Edinburgh, and Henry G. Plimmer, a lecturer and pathologist at St Mary's Hospital, London. Russell's and Plimmer's bodies were the subject of many papers, which included drawings of magnified tumour sections along with detailed instructions on how to prepare and stain the sections with dye in order to reveal the relevant details.[4] Publications often also included warnings that much practice in the art was required before finding the parasite would become routine. Criticism of these findings was intense, with other researchers declaring the structures to be merely products of cellular degeneration, stray blood corpuscles, or normal cellular elements with an affinity for particular dyes. In the early twentieth century in Great Britain, hopes that a parasite might be discovered began to fade, with the Imperial Cancer Research Fund (ICRF) declaring the search futile in 1905.

One reason the germ theory of cancer persisted in attracting a handful of researchers and patrons was that it suggested vaccines against cancer might be found. In the USA, this hope was kept alive by Dr Roswell Park, Professor of Surgery at the University of Buffalo School of Medicine, who established the New York State Pathological Laboratory in 1898. Although the centre covered all aspects of cancer treatment and research, Park's strong conviction that a parasite was responsible for cancer meant that the search for vaccines was an institutional priority. Dr Harvey R. Gaylord, also an ardent supporter of the germ theory, succeeded Park as Director in 1904; Gaylord had claimed in a paper that Russell's and Plimmer's bodies displayed a direct morphological link and were therefore different life stages of a single parasite (*The Practitioner*, 1899).[5]

Nonetheless, the number of orthodox clinicians researching parasitic or bacterial causes for cancers dropped considerably in the early twentieth century, in the USA as in Great Britain. Influential cancer specialist, James Ewing, published attacks on the theory of parasitic causation from 1919, and sponsorship of this branch of experimental work by the United States Public Health Laboratories came to an end when the established cancer experts of the day declared the research to be of an insufficiently high standard, even suggesting samples of cancerous tissue that had yielded bacteria had been contaminated through poor working practices (Falcone, 2005).

However, cancer viruses were shown to be unquestionably real. In 1910, Peyton Rous demonstrated that a number of cancers found in chickens could be passed from bird to bird by injecting extracts of tumour tissue that had been previously filtered to ensure they did not contain cells or bacteria; at this time, viruses were known as 'filterable agents', too small to be seen by any existing instruments (Rous, 1910, 1911; Van Epps, 2005). This filterable agent of chicken sarcoma was not disputed in the way that claims about cancer parasites had been, but it remained an isolated finding until the 1930s, when researchers at the Rockefeller Institute identified another carcinogenic virus, this time affecting rabbits (Gaudillière, 2006).

The relation between cancer and viruses was pursued in cancer research centres in Great Britain and the USA. This research intensified in the postwar period, as new techniques for isolating tiny molecules from tissue samples, and for making these visible, transformed viruses into popular research objects.[6] Viral research techniques expanded the power and breadth of the toolkit available to *all* molecular biology researchers, as demonstrated by Angela Creager (2001) in her history of the significance and impact of studies of the tobacco mosaic virus at the Rockefeller Institute during this period. It was the work of Ludwik Gross during 1953, however, that excited scientists at the NCI with the possibilities it presented for cancer research and cancer control. His demonstration that filtered extracts of tissues taken from mice with leukaemia could initiate cancer in other animals prompted many attempts to replicate his findings; over the following four years, seven other American scientists identified further cancer-causing viruses in animals, and tumour virology became an attractive field for research (Baker, 2004, pp. 9–13; Javier and Butel, 2008).

In 1958, the United States Congress made additional funds available to the NCI to instigate a large-scale research programme to explore tumour virus biology systematically. A new grants panel was established, charged in its first year with distributing a million dollars to researchers; two years later this figure had risen to nearly four million and by 1964 to more than ten million (Baker, 2004, pp. 14, 22, 114).[7] As Jean-Paul Gaudillière has shown, much of the NCI-funded early work focused on ways of standardising procedures and improving tissue culture techniques, thereby reducing the risk of losing cell lines and contaminating samples, and increasing the reliability of results. By 1964, a network of laboratories had been established, a set of agreed standards was in place, and the Special Virus Leukemia Program was launched. As the search for viruses capable of causing human cancer intensified, nonhuman primates became the test organism into which filtered tissue and extracted viruses were injected (ibid, p. 113). The number of test animals required increased dramatically, but there was confidence these efforts would produce effective cancer prevention techniques.

The NCI was not the only body funding such research. In Great Britain, from the late 1950s the MRC and the two major cancer charities – the British Empire Cancer Campaign (BECC) and the ICRF – were each investing in laboratories testing for viruses in tumour samples, and screening those viruses already known to have carcinogenic properties, in animals and in people. Cancer immunology was also seen as a promising area for research funding during the 1960s in France, where Georges Mathé and colleagues gave patients with leukaemia combinations of BCG (Bacillus Calmette–Guérin) and tumour-specific vaccines to stimulate their natural defences. As we shall see, Burkitt's discovery of the first human cancer to apparently be caused by a virus came at a time when faith that such patterns of causation would be found, if only sufficient money and talent were set to the challenge of seeking these tiniest of carcinogens, was at its peak.

The theory that *bacteria* might cause cancer had been waning since the 1920s; in the 1950s there were very few clinicians that still took the possibility seriously. The theory's proponents, however, were voluble and thus visible to families affected by cancer and to the journalists who wrote about them, and thus news stories about 'cancer germs' rarely distinguished between viruses and bacteria (Cantwell, 2003).[8] Historians who have studied the public health campaigns of former periods have noted that many people in the 1960s did not understand the difference between bacteria and viruses; given doctors' reports of requests from their patients for antibiotics to treat viral infections, it could be argued this is still the case today.

That highly esteemed, published medical practitioners and researchers had been publicly speaking about cancer being caused by infectious agents for many years had a profound impact on the interest taken in Burkitt's work, both in the biomedical world and in the media. Discoveries that implicated microbes in the development of cancer in humans or other animals fuelled the hope that vaccine therapy might be of value in treating cancers.[9] Immunology, a relatively new field, was seen to hold enormous potential, suggesting that the 'immune system' – the body's own healer – might be strengthened through medical intervention.[10] But the greater promise, one whispered in the news throughout the 1960s, was that vaccines might *prevent* cancer from striking in the first place (see, e.g., *The Times*, 1964a).

The Tumour Safari

Denis Burkitt was a devout Christian who felt called to serve as a doctor. Though he aspired to work as a missionary doctor, thereby following in the footsteps of his hero David Livingstone, his attempts to find a placement in Africa were initially unsuccessful (Epstein, 2004).[11] He was rejected on the grounds of his poor eyesight: Burkitt lost his right eye in childhood, though this appears to have neither affected his surgical skills nor his powers of observation. His second application to the colonial service was successful, and he began work as a general surgeon in Mulago Hospital, attached to the medical school of Makerere University, Kampala, Uganda, in 1948, where he grew especially interested in designing prostheses from locally available materials for use following operations to correct congenital malformations.

Burkitt's work changed direction in 1957. He was called to consult on a young boy in one of the wards at Mulago Hospital who had unusual swellings on the top and bottom halves of both jaws. This was not something Burkitt had witnessed before, but shortly afterwards, while carrying out surgical sessions at a different Ugandan hospital, he met another child with the same pattern of growths. Burkitt recalled that he 'plonked the mother and the child into my car and I drove them back to my wards in Mulago Hospital and began to investigate'. This second case of an apparently rare syndrome led him to wonder how many other cases of this distinctive tumour growth pattern were being seen by doctors in his part of the world.[12]

Burkitt began to amass records of all the cases seen in children at Mulago Hospital, and made contact with other government and missionary hospitals across Africa in order to map the condition's incidence. The tumour presented in diverse forms and had hitherto been classified by site of largest mass. However, Burkitt was convinced that cases in which children presented with multiple and rapidly growing tumours, particularly in the jaws, were better classified together as a new and very different condition, one with multiple 'primaries' rather than one initial mass with metastases. The hospital pathologist, Dennis Wright, confirmed that cells seen in these cases of cancer were of the same kind in different children, whether the tumours were located in the jaws or in the abdomen, and resembled those seen in familiar varieties of lymphoma. Burkitt had, indeed, identified a new type of cancer.

In 1958, Burkitt secured a small grant from the British government for the production and distribution of a poster illustrating two children with multiple jaw tumours – seen in around half the patients with this disease pattern – asking for information from other clinics in which the tumour had been observed.[13] Over the next three years, Burkitt marked with hand-coloured drawing pins more than 300 cases on a map in his office, and kept files on patients from numerous countries in East and tropical Africa, demonstrating the cancer was unknown in some regions while comparatively common in others.[14] As the population of Westerners in Uganda was small, Western staff of Mulago Hospital frequently shared meals and social events with the scientists employed at a nearby government biomedical research laboratory, the East African Virus Research Institute in Entebbe. At such an event, Burkitt discussed the incidence of his tumour with an entomologist studying insect-borne viruses, Alexander John Haddow.[15] It appeared that Burkitt's map of cancer distribution was remarkably similar to maps of insect-borne viral illnesses such as malaria and yellow fever, both known to be carried by mosquitoes and thus limited in incidence to those regions where temperature, humidity, and altitude suited both fly and infectious agent. However, as the map was based on a small sample of cases reported by doctors scattered over a wide region, it did not represent absolute numbers of cases per million of population and was not of a sufficiently scientific nature to permit any definite conclusions about the aetiology of the tumour. Further work was needed to gain a more complete picture of incidence and explore which environmental factors were critical.

The fact that A. J. Haddow was working in a team funded by the ICRF was to lead to many disputes over who should bear the cost of supporting Burkitt's research, given that several funding bodies were already supporting related research. A. J. Haddow and Sir Harold Himsworth, secretary of the MRC, appear to have been quite pragmatic in their dealings with one another: what mattered was that *someone* paid so that the work could be *done*. A. J. Haddow was clear in his dealings with Himsworth about the significance of Burkitt's findings, despite the fact that all evidence accumulated to date was anecdotal. A. J. Haddow took the clarity of the early incidence map to be an indication of the strength of the viral hypothesis:

if by such crude and empirical methods one can arrive at a reasonable map of the tumour distribution then that distribution is almost certainly a true ecological one. When Burkitt returns from leave he hopes to fill some of the gaps in his map, and he and I plan a much more thorough attack on the climate-cum-zoology angle.[16]

Burkitt (1958) announced his findings in a British journal article, but one with a small circulation, read by few cancer researchers. Although this initial publication did not receive much attention, knowledge of his work spread by word of mouth, through the connections of the Virus Research Institute scientists, the contacts Burkitt had made through the MRC and the British cancer charities, and the NCI visitors he had received. That this cancer might be induced by a virus was enormously exciting for those researchers being funded by the NCI's new tumour virus panel, and those funded by the three chief British medical research organisations. If proven, Burkitt's discovery would justify the vast amounts being spent on experiments with flies, mice, and monkeys, which had as yet yielded no benefit for any cancer patient.

Burkitt began receiving medical visitors from the USA, Great Britain, and other African nations. In January 1960, leading researchers and clinicians from the Sloan Kettering Institute (SKI), including Joseph Burchenal, travelled to Nairobi in order to test new drugs on cancer patients there. Because there were no radiotherapy facilities between Cairo and Johannesburg, central Africa afforded fantastic opportunities for chemotherapy research: as patients had not received radiotherapy, results were easier to interpret, and the ethical dilemma of testing new substances on people was easier to negotiate in the absence of any alternative possible treatment. While in Nairobi, Burchenal made a detour to Kampala to visit Burkitt and witness the exciting work being conducted there and at the Entebbe virology institute; he left behind some methotrexate pills. A second scientist from the Institute, human tumour immunologist Herbert Oettgen, visited later to give instructions on how to use them.

By January 1961, Burkitt was planning to increase the scale of his epidemiological endeavours, and drafting funding applications to mount a 'tumour safari' to plot the edge of the 'tumour belt'.[17] The notion of an illness safari strikes the modern ear as strange, as the term is now exclusively associated with trips to Africa to shoot or visually capture rare large mammals in their native habitats. But the word 'safari' literally means any overland journey, particularly those undertaken for hunting or exploring purposes, and originates from the Swahili word for 'a long journey'. Historically, the term has been closely linked to trips within Africa, in part because of the roots of the term, in part owing to the writings of American author Ernest Hemingway, seen as the person responsible for the popularisation of the term in the English language following his first hunting trip to Africa in 1933, though its use predates the novel he wrote following that experience.[18] The word's association with hunting also appears to derive from Hemingway, but in Swahili the term connotes discovery or adventure rather than the acquisition of trophies (Paice, 1999). Illness safaris should not, then, be understood as morbid searches to capture rare or wild diseases. Rather, Burkitt's safaris, like Livingstone's trips, were designed to discover what was going on in Africa's interior: to disprove Western assumptions about the nature of life in the continent.

Funds for the safari were drawn from at least three sources, after a long letter-writing campaign seeking support, and promises that every funder would receive due credit in future publications. The BECC donated £100 towards the purchase of an old Ford station wagon to transport Burkitt and two colleagues on their journey, mapped to cross and re-cross where they thought the edges of the 'tumour belt' would be. Similarly, the SKI made £100 available to Burkitt for the trip's expenses.[19] Burkitt calculated that the car alone would cost at least £250. He discussed with Himsworth the possibility of the MRC supporting the venture, while Himsworth was in Uganda early in 1961 visiting A. J. Haddow. Himsworth used his influence to secure funds quickly so that he could promise them to Burkitt before the three doctors were due to embark on their travels.

A decision was made that the BECC and the MRC between them would not only send money for the safari itself, but would fund Burkitt's project with £5,000 over the next four years to pay for a secretary based in Kampala to manage the compilation of data. The MRC also pledged the assistance of its epidemiologists. As the two organisations dithered with each other and with the ICRF over who would pick up the tab, there was no shortage of other offers of support. The director of the Leverhulme Trust, Sir Miles Clifford, wrote to Himsworth offering to step in and make funds available to the MRC in order to end the deadlock in negotiations between the BECC and ICRF over who should fund what and get the work underway.[20] The wrangling came to an end in August when Himsworth hinted to the Colonial Office that if the British organisations delayed any longer, the Americans would take over the project and steal the credit:

We have just heard from Kampala the not unexpected news that although they would like this to be a British venture private American foundations are now pressing them hard to accept workers and funds.[21]

The MRC served as a broker between all interested parties: Himsworth established a working party to coordinate and conduct all research strands which would then be jointly managed by the two competing charities, represented for the ICRF by Dr G. F. Marrian (Chairman of the Research Advisory Committee) and Dr R. J. C. Harris (Division of Experimental Biology and Virology), and for the BECC by Professor Theo Crawford and Tony Epstein, with A. J. Haddow as the Entebbe representative. Both charities indicated they would ultimately split the work costs, but as the committees were taking too long to process the proposals, Himsworth went over their heads and secured funds from the Colonial Office's pot under the MRC's control. Although London aspired to be the centre of Burkitt's African network into which dozens of small rural hospitals reported, it was Kampala that drew the visitors and scientists, funders and journalists. Far from being a peripheral site in a network of grand cancer research centres, Kampala became a hub for research into human tumour virology, with London sending resources and representatives to Uganda as well as receiving them.[22]

By the start of 1961, Burkitt was already famous thanks to his network of personal contacts across Africa and in American and British cancer research centres. Interest snowballed that year, however, following the publication in *Cancer* of a paper on the syndrome Burkitt had observed, alongside another by American pathologist, Gregory O'Conor, who worked with Burkitt in the medical school in Kampala (Burkitt and O'Conor, 1961; O'Conor, 1961). These caught the attention of a wider circle of cancer researchers in the USA and of scientists working on MRC grants. Importantly, they also drew the interest of a number of journalists, who wrote about the discovery and its significance to the cancer problem as a whole, thereby placing Burkitt's work in the public eye for the first time. In discussion with one of his biographers, Burkitt stated that from this point on, he was inundated with requests for copies of his papers, for more information, and for samples of tissue collected from his patients:

After these papers came out ... people said, *This is something*. There were leaders in *The New York Times*, and so on; and we were flooded with requests for reprints (Glemser, 1971, p. 107).[23]

Burkitt's work was more widely reported in American newspapers and popular magazines than in the British press, in keeping with a general reticence seen in British reporting of cancer, especially of cancer affecting children, during this period.

Mapping Mosquitoes

Burkitt's journey in the last months of 1961 took him and his colleagues, Ted Williams and Cliff Nelson, to more than 50 hospitals in Eastern and Southern Africa. Ted Williams was the director of a mission hospital near Arua in northeast Uganda; Cliff Nelson was a Canadian who had been in government medical service in Uganda and whose contract had just expired (ibid, p. 73). They spent months making preparations, customising the car for rough terrain and writing to the hospitals they wished to visit with expected dates of arrival. They also prepared a book of information on the tumour to illustrate what they were hunting. In recollections recorded in 1991, Burkitt explained how the men gathered their geographical evidence:

> We made an album of photographs showing clinical pictures, X-rays, pathology, so that we could sit down … and turn the pages over and say to the doctor, 'Have you seen this or have you not?' And he would say, 'I remember a girl or something'. And we'd say, 'Where did she come from?' and we would put a little mark on the map, that's the way we did it.[24]

The three men covered more than 50,000 miles in the course of their ten-week safari. They were able to gather information on many new cases of the cancer, some currently undergoing treatment or recorded in doctors' files, others just in the memories of long-serving clinicians. Burkitt found those doctors working in small mission hospitals in poor areas frequently more useful than government-appointed doctors and district medical officers because they had remained on the clinical side of medicine and stayed in one place for so long, in many instances for decades. Burkitt's Christianity, and the fact he was travelling with two missionary doctors, made his appeals for cooperation to these overworked medics all the more powerful: he spoke the same language of faith, and demonstrated through his choice of travelling companion that he valued their type of expertise, as Ted Williams described, when interviewed by one of Burkitt's biographers in the late 1960s:

What he [Burkitt] did, in fact, was to alter the whole concept of research. He showed that clinicians, particularly people who work for a long time in one place – as we missionaries do – are in a position to assemble a great deal of unique and extremely valuable information (Glemser, 1971, p. 103).

The following year, Burkitt completed his mapping by flying to the Eastern states, areas he had been unable to reach in the long safari owing to excessive rainfall having washed away the roads. The additional findings from these subsequent shorter safaris demonstrated that annual rainfall was also vital in explaining the tumour's pattern of incidence: this conclusive final factor led A. J. Haddow to stand publicly by his hypothesis that the tumour was caused by a virus transmitted by insects.

Powerful Little Pills

Burkitt's epidemiology and illness safaris established Mulago as the centre collecting scientific data from government and missionary hospitals. Burkitt took steps to guarantee ongoing supplies of drugs for clinicians prepared to cooperate in his efforts to map incidence and uncover aetiology, recognising that the possibility of treatment would tempt more physicians and surgeons to put in the hours necessary to supply him with data. Lederle, the pharmaceutical company whose collaboration with Sydney Farber we have already discussed, supplied drugs free of charge, on precisely the grounds that trials on untreated African cancer sufferers were easier to analyse and justify than those on American or European patients.

By 20 August 1962, Burkitt had tested methotrexate on 20 patients and the results were being analysed by Oettgen at the SKI in New York.[25] By May 1963, Burkitt reported to the MRC, funders of his epidemiological work, that he had administered chemotherapy (mercaptopurine, methotrexate, and Endoxan) to 75 patients, of which 51 responded and nine remained well. This included 21 children given methotrexate, of which ten responded and two were apparently cured.[26] Through such demonstrations that his was a tumour 'particularly suitable for the assessment of cancer chemotherapy', Burkitt secured free drugs and ongoing funding.[27]

These clear proofs of chemotherapy's efficacy against cancer were remarkable: at the time of Burkitt's results, only choriocarcinoma, a very rare cancer of the placenta, had been shown to be curable through chemotherapy. Burkitt's success, therefore, limited as it was, drew enormous media interest in the USA: a television documentary was filmed in 1962, popular science magazines published lengthy pieces about his research and its connections with that of virologists in a nearby institute, and Burkitt was sought for interviews by newspaper and broadcast journalists whenever he attended cancer meetings in North America.[28] By contrast, magazine and newspaper coverage of his work in Britain during 1962 and 1963 focused solely on the scientific potential of his work: how it linked with findings emerging from other laboratories about the role of viruses in the growth of cancerous tumours.[29] Burkitt's name was rarely mentioned in newspaper articles, though the man behind the 'Impressive ... story from Uganda', and his colleagues in the region, received some page space in longer articles in general science publications.[30] American stories tended to focus on the human side of the story – the heroic explorer doctors saving sick children – while the children affected by the lymphoma barely featured in British news stories at all. This might be because very little 'marketing' of the story was undertaken by the organisations that funded Burkitt's work. This difference in national attitudes will be explored in greater detail in Chapter 5.

The demonstrated utility of anticancer drugs against two aggressive solid tumours was presented in the news coverage, in the USA and in Great Britain, as proof that the vast investment in the NCI's chemotherapy programme and support for chemotherapy research in Great Britain were justified: cures were being discovered for cancers that would otherwise quickly kill every sufferer. Choriocarcinoma affected a tiny proportion of women who had recently undergone pregnancy, and was invariably fatal until 1956, when a small group of scientists at the NCI announced that methotrexate rendered it curable for a substantial proportion of patients.[31] That this cure had been developed within the first years of the NCI's chemotherapy programme instilled optimism within the teams working with new drugs available for clinical trials. That Burkitt's tumour succumbed to drugs just a few years later increased the NCI's confidence that even childhood acute leukaemia would eventually be cured through drugs: remissions would become permanent. These successes were presented in medical publications and the popular press as the first of many, heralds of breakthroughs against much larger cancer problems. Burkitt's success was an additional boon for all those committed to seeking the future of cancer control by chemical means because this disease appeared to have similarities with other childhood cancers closer to home.

Burkitt's success was made all the more moving by the fact that his patients were almost always young children from families facing major adversity in a region where malaria was endemic, and child mortality from diseases unheard of in the developed world was high. That Western biomedical research could save or at least prolong the lives of these most unfortunate – and arguably picturesque – cancer patients, played well as both a human interest story in the news and as rhetoric in appeals for research funding to governments and private donors. That Burkitt was able to treat children so cheaply, without the hugely expensive beds being prepared for chemotherapy patients in leading research hospitals in the USA and Great Britain, was also emphasised by those pushing chemotherapy as the future of cancer treatment.[32]

But many medical experts disapproved both of the way in which Burkitt had treated patients and of the technologies he used to record results. Owing to lack of facilities for supportive care, Burkitt could not give treatment to the limits of toxicity, as was the orthodoxy at that time amongst chemotherapy researchers in the USA. As families occasionally removed their children after just a few days in hospital, he was also unable to prescribe long courses of pills or even establish results for many patients. In 1966, he published results of his chemotherapy tests on children suffering from what was by then known as 'Burkitt's tumour' in which he stated that of 90 patients treated, 38 were 'lost': it had not been possible to follow up with the families to find out how effective treatments had been. Instead of representing survival as a percentage, and risk of relapse on a survival curve measuring number alive over time, he submitted for publication a photo of his 12 known survivors, lined up along a wall outside his hospital. As there was no scientific rationale by which to decide on the boundaries of some group from which the 12 could be seen as a proportion, he presented absolute numbers of survivors rather than percentages (Burkitt, 1966).[33]

Burkitt kept detailed records of all the children he treated for the lymphoma, assigning each a number after the letter 'J'. His published results of successful treatment gave the presenting symptoms of each of the patients, an account of the medication administered, and a history of the follow-up appointments attended (in the Mulago hospital or in patients' homes) noting any residual problems experienced by the children. The photograph of the survivors was annotated, linking ten of the faces to their corresponding case numbers: from left to right, they were J174, J167, J37, not numbered, J76, J176, J289, not numbered, J95, J303, J180, and J113. Burchenal reprinted the photograph in an article the same year that suggested that the African lymphoma could be seen as a 'stalking

Figure 4.1   Denis Burkitt's 12 known survivors, published in 1966
Source: Reproduced by permission of the Wellcome Library, London, and the Burkitt family.

horse' for acute leukaemia, as research into the cancer cells showed the two types were pathologically very similar (Burchenal, 1966).

This group of survivors had been gathered together on the occasion of a conference. Such was the interest in Burkitt's lymphoma and the research being undertaken by others into its causes and treatments, the Union Internationale Contre le Cancer (UICC) chose the cancer as its theme for the 1966 International Cancer Congress, held in Kampala in January. The UICC had been established in 1933 to foster international cooperation and sharing of results in the effort to understand and control cancer; hosting a meeting meant mixing with the world's leading figures in one's particular branch of cancer research. At this meeting, Burkitt's findings and their applications were discussed by James Holland, Emil Frei, and Howard Skipper – three of the leading leukaemia researchers then working in the USA – to an audience of over 30 investigators drawn from 12 countries. Burkitt's own paper on his use of low doses of chemotherapy caused a stir by including a presentation of this group of living survivors.[34]

From 1962 to 1966, the President of the UICC was Professor Sir Alexander Haddow, Britain's leading chemotherapy researcher. As such, it fell to Haddow to close the conference with a short speech to summarise the key developments shared at the meeting. Haddow chose this moment of presentation as his theme; he spoke eloquently about the impact of seeing a procession of 12 surviving children in the conference hall:

> No one who witnessed it can ever forget yesterday's demonstration of Burkitt's patients – cured African children, we hope. It was an experience expressively moving and of the utmost practical significance, not only for these children and their families, but also for us in proving that gross pathology can be reversed by purely chemical means (Burchenal and Burkitt, 1966).

While other papers at the conference appealed to reason with tables of results or suggestions for increasing the cell kill rates and thus survival times, Burkitt reached the heart with his dozen walking results.

Finding Cancer Microbes

In the March of 1961, Burkitt had spent some weeks in England, visiting potential funders, friends, and family members, and giving a few lectures in British medical schools. A regular visitor at the Middlesex Hospital, Burkitt often entertained students with slides of tropical diseases never seen in London. On this visit, however, Burkitt took as his topic the tumour syndrome he was then preparing to map. One of the audience members was Tony Epstein, a virologist who had been working on Rous' chicken sarcoma. Epstein's curiosity was piqued by the geographical pattern Burkitt described and was seeking to further establish through his safari. The map Burkitt revealed suggested a tumour that depended on both altitude and latitude, appearing only below 5,000 feet at the equator, below 3,000 feet once 1,000 miles south of the equator, and below 1,000 feet a further thousand miles away from the mid-line. In other words, the cancer depended on temperature, strongly suggesting a viral factor. Following the lecture, Epstein introduced himself and the two men established a longstanding correspondence and friendship.[35]

Despite an agreement that tumour samples would be flown from Kampala to London for Epstein to analyse, no shipments successfully made the journey for six months while there were negotiations over how samples would be carried. BECC funding enabled Epstein to travel to Kampala that September and meet the pathologists handling samples from Burkitt's patients, Dennis Wright and Greg O'Conor.[36]

Only after this face-to-face encounter and further political wrangling back in London between the agencies keen to fund the work but also limit their expenses, did cancerous tissue begin to travel from Uganda to Great Britain. Wright oversaw the dispatch of pieces of each tumour treated in Kampala to Epstein in London, via British Airways flights and personal pick-ups by members of Epstein's team at Heathrow.

Burkitt's patients were the source for tissue samples sent to hospital laboratory researchers in London, New York, and Toronto, all seeking to isolate the virus or viruses responsible for the tumour, and demonstrate causation, not mere correlation. Not only was it difficult to arrange reliable and temperature-controlled transport for biopsy materials from Uganda to Great Britain and the USA, but it was even harder to establish living tissue cultures from the samples. In 1963, Tony Epstein and his colleague Yvonne Barr, funded by the NCI at the Bland Sutton Institute attached to London's Middlesex Hospital, isolated a herpes-like virus from the culture grown from one patient's samples. They soon succeeded in identifying what appeared to be the same virus in other samples (Glemser, 1971, p. 173). However, they were unable to establish what illness this virus caused in those affected who did not develop the rare tumours. They were also unable to state categorically that its presence in tumour materials indicated any *causal* role in the development of cancer.

In 1964, Epstein and Barr published on their discovery of the new virus, subsequently named after them, in *The Lancet*, but its role in the development of the African lymphoma was far from clear. Two days after isolating the virus, Epstein sent samples to a husband and wife team working in Philadelphia, Gertrude and Werner Henle. These virus researchers were well known to Dr Everett Koop, the Professor of Paediatric Surgery heading the oncology surgical team at the Children's Hospital in Philadelphia. His keen interest in cancer research, especially the potential of chemotherapy to help children suffering from metastatic disease beyond his reach, had prompted Koop to travel to Uganda to meet Burkitt in 1962. That year, Koop had suggested to the Henles that Burkitt's tumour could conceivably become the first case of human cancer with a demonstrable viral origin. The Henles were therefore very keen to collaborate, once Epstein had material they could analyse. Having confirmed it was, indeed, a new herpes-like virus, the Henles went on to discover that 85 per cent of people, from all over the globe, carried antibodies to it (ibid, pp. 177–8). Every tissue sample from a patient suffering from Burkitt's tumour contained examples of this distinctive virus.

Given its prevalence in the population at large, the mere presence of the virus could not be enough to trigger the development of a tumour. Research scientists from NCI-sponsored projects across the USA, Canada, and Great Britain, as well as those in African countries, turned their attention to discovering what other illnesses the virus could cause, and what other factors were needed to stimulate cancerous growth. A laboratory technician employed by the Henles, Elaine Hutkin, who did not have antibodies to the Epstein–Barr virus, donated blood to serve as a culture medium for use in their experimental work. Elaine contracted a bad infection with symptoms matching those ascribed to infectious mononucleosis, otherwise known as glandular fever, the aetiology of which had not at that time been established. Upon her recovery, her blood was found to contain antibodies to the Epstein–Barr virus: could it be that Burkitt's tumour was an unusual reaction to the agent that caused glandular fever? By accessing the medical records of Yale University, the Henles were able to test whether Hutkin's illness had truly been infectious mononucleosis. Between 1958 and 1964, Yale had collected blood samples from all their new students as part of a study on the illness conducted by Dr James C. Niederman. Cross-matching frozen serum from a selection of students, with records of who had and who had not suffered from infectious mononucleosis, revealed that every student who had had the illness carried antibodies to the Epstein–Barr virus (ibid, pp. 181–2). The cause of glandular fever had been identified. However, the causal role of the virus in Burkitt's tumour required further research.

Gilbert Dalldorf provided the missing piece of the puzzle. Dalldorf was a highly respected virus researcher at SKI who had played a prominent role in polio research. In 1962 he published his hypothesis that the virus could only cause Burkitt's tumour if the sufferer's immune system had already been compromised through exposure to severe or repeated infections, or had otherwise been damaged. Over the next four years, Haddow, Burkitt, Epstein, Dalldorf, and others put the pieces together and established the most likely aetiological route: that the Epstein–Barr virus could trigger cancerous changes in reticuloendothelial cells in children whose immune systems functioned poorly owing to hyper-endemic malaria. The final piece of evidence suggesting the condition resulted from the virus *and* a damaged immune system came from epidemiological studies conducted in the mid-1960s, demonstrating the tumour was unknown in two regions that met the conditions for temperature and rainfall, but which had been the subject of successful malaria eradication programmes: the islands of Pemba and Zanzibar off the coast of Tanzania, and the area around Kinshasa in the Congo (Burkitt, 1967).[37] Since 1970, this account of the role of viruses in the development of Burkitt's lymphoma has largely remained stable.

Faith that the immune system could be supported to fight off cancer grew in the 1970s, as cancer immunologists promoted their developing therapies as a fourth modality of cancer care. In his 1978 book, *Man, Cancer and Immunity*, Glasgow pathologist Alistair Cochran described his amazement at the 'ingenuity' involved in the 'veritable cornucopia' of immunotherapeutical approaches, and expressed little doubt that at least a few would be successful. Immune system imagery entered the popular imagination in the 1980s, as described by Ilana Löwy (1994, 2001), once electron microscopic photographs of killer cells attacking a tumour cell were published appearing to show the body as a battlefield where the immune system defeated the 'enemy from within'. Burkitt's work with the African lymphoma served as a timely confirmation that this vision of cancer and the body's role in resisting mutations held biological value (Cochran, 1978; Haraway, 1993; Barnes, 2006).

Burkitt's Tumour Conceptualised as an Exemplar Childhood Cancer

By 1966, research by Burkitt, Epstein, the Henles, Dalldorf, and scientists at SKI was indicating that Burkitt's tumour was closely related to acute lymphoblastic leukaemia. Clinicians, including Burkitt and Burchenal, believed that improved understanding of acute lymphoblastic leukaemia would come from closely studying the natural history of Burkitt's tumour and how this was altered by different treatments. Burkitt's tumour responded strongly and rapidly to chemotherapy, relaying hope to those working with leukaemia and to sufferers, and its apparent viral aetiology suggested the NCI's investment in virology was justified.

Burkitt's tumour was frequently discussed in lectures and publications about leukaemia. In 1963, the Leukaemia Research Fund (LRF) began funding and publishing a series of annual guest lectures on leukaemia. Professor Jean Bernard, a French pioneer of treatment for malignant blood diseases, gave the first. The second was by Burchenal. Both mentioned Burkitt's work (Bernard, 1964; Burchenal, 1965). Burkitt himself delivered the third in 1966, in which he stressed the relevance of the tumour he had discovered for researchers seeking solutions to leukaemia. That these two cancers belonged together was repeated in journal articles and popular science magazine pieces, tying them together within the growing field of specialism in childhood cancer.

At the 1966 UICC conference in Uganda, Burkitt and Burchenal explained that both Burkitt's tumour and acute lymphoblastic leukaemia were types of cancerous mutation taking place in reticuloendothelial cells, and presented the two cancers as being different 'host' reactions to such mutation. In his 1966 LRF lecture, Burkitt further raised the possibility that the cancers might be the result of different stimuli affecting the same tissue, or the response to a common stimulus resulting in different clinical manifestations (Burkitt, 1967, p. 5). Burkitt's tumour appeared to be a localised cancer, and if tumours regressed in response to chemotherapy there was a good chance there would be no recurrence: the patient would be cured. Acute lymphoblastic leukaemia, by contrast, was a generalised cancer that almost always recurred, even following chemotherapy-induced remission. Burkitt concluded his lecture by outlining the tantalising implications of the differences and similarities:

> Could it be that a strong antibody response in the case of African lymphoma results in localised lesions, whereas a poor response in the case of leukaemia allows a generalised malignancy (Burchenal, *personal communication*)? Is there a reciprocal relationship between the two conditions, as has been suggested independently by Dalldorf (1962) and O'Conor (O'Conor and Davies, 1960)? ... Could a reticuloendothelial system altered by repeated insults from malaria and other infections suppress a generalised leukaemia infection, allowing only local tumour deposits? (Dalldorf *et al*, 1964; Edington, 1964)? (1967, pp. 22–3)[38]

Burkitt's tumour also served as an exemplary children's cancer because Burkitt himself pursued and presented his research in a personal manner, utilising his close connections with missionary doctors to track down cases of the tumour in rural communities across Africa, and including descriptions and photographs of the children he treated in his publications and presentations. Burkitt's studies drew media interest both because of their apparent relevance to cancer research more generally, but also because they afforded an opportunity to present a human side of cancer research, to tell stories of committed doctors and desperately ill children to which a broader public might be expected to relate. We cannot discount the additional attraction afforded by what could be called the National Geographic Effect: this was picturesque research, with white doctors sweating in the tropics to save the lives of impoverished black children, demonstrating true Christian charity and a legacy of colonial rule.

Thirty years later, Burkitt attempted to explain the appeal of his research, the surge in interest shown by journalists and their readers and viewers in the early 1960s:

> when people, the ordinary layman, thinks of cancer research, he doesn't think of crossing rivers on ferries and all the rest of it; he thinks of albino mice in the laboratory and white coats ... It never occurred to us that it was going to come out in the *Reader's Digest* later on and so on, in the sort of language of Burkitt and Ted and Cliff turning their back on comfort and everything for the sake of suffering humanity.[39]

Burkitt believed that public interest flowed from the humanitarian nature of their travels, and from the lack of obvious science in what they were doing. Burkitt seemed to demonstrate that cancer research could be the opposite of high-tech: breakthroughs could flow simply from good powers of observation and a determination to do good for others. There is something missing from this account, however: public interest was already focused on childhood cancer in the 1960s. In the next chapter we discuss the reasons behind the increase in coverage of childhood cancer in Great Britain during the 1960s, and the impact this had on researchers, and on patients and their families.

5
Making the News and the Need for Hope

From the 1930s onwards, the American media carried human interest stories about families touched by leukaemia or childhood solid tumours. In stark contrast, British newspapers were practically silent on the subject until the 1960s.[1] Why were such stories absent from the British press for so long? Was it just the fabled British reticence, or were the levels of coverage in some sense deliberate and negotiated? In this chapter we explore the reasons behind the entry of childhood cancer into the news, analysing what changes led to stories of medical breakthroughs, setbacks, and of children's battles for life being deemed to be in the public interest. We shall see that the self-promotion of charities was central to the rising interest in childhood cancer: children's cancers attracted more public donations than work on cancer affecting adults, and stories about research generally proved more effective than accounts of treatment. Laboratory research on the role of viruses in the development of certain malignancies became more newsworthy with reports of the Ugandan work on the African lymphoma, as discussed in Chapter 4. The public thirst for progress in the war on cancer also brought into the spotlight the plight of 'hopeless' cases, who, having exhausted all mainstream options, often resorted to alternative remedies. We will examine the media interest in one such family's quest, and reactions from the orthodox medical establishment, in the period just before biomedical treatments began to secure small advances.

American Media Interest in Childhood Cancer

As Gretchen Krueger (2008) has shown, media coverage of children with cancer was a carefully planned development in the USA, designed to solicit public interest and donations towards the costs of research and purpose-built accommodation. Stories of families' battles with long odds of survival and tough treatment choices had featured in the leading newspapers from the 1930s, but the tone of the coverage of childhood cancer shifted and became more hopeful in the 1940s, in response to breakthroughs in treatment and in line with changing public information campaign approaches as deployed by charities raising money for polio research. The Jimmy Fund, established in 1948 to support the work of Sydney Farber at the Children's Hospital in Boston, explicitly replicated the strategy that had proved so successful in generating support for the March of Dimes just two years earlier. The March of Dimes was a fundraising campaign, launched by the National Foundation for Infantile Paralysis in 1938, which asked every individual to contribute a small amount of change towards the research effort seeking a vaccine (Helfand et al., 2001). In 1946, The National Foundation began a process of selecting children from different regions in the country as images for posters, composing visual stories that compared each child's condition before and after treatment paid for by the Foundation. The Jimmy Fund chose to follow the technique of focusing calls for donations around individual children and their experiences of sickness and hospital treatments. Within a few years, the American Cancer Society adopted the strategy, featuring children in its campaign literature and rallies to capitalise on the public's enthusiasm for investing in the hopes of child cancer patients (Patterson, 1991). Cancer charities promoted their endeavours to newspapers and current affairs weekly and monthly magazines, resulting in photo-rich articles showcasing the work of particular hospital research and nursing teams, and how this benefitted a number of spotlighted child patients; these charities ensured the plight of children fighting cancer remained in the public's eye, always with the message that donations could make the difference between death and cure (Krueger, 2008).

   In Great Britain at this time, newspapers did not cover individual families' efforts to help their children win out against cancer. Great Ormond Street Hospital did make use of photographs of children being cared for in its wards in fundraising campaigns from 1946, perhaps consciously copying the effective strategy of the March of Dimes, but the campaign centred on the hospital as an establishment to help the sickest children, and not on specific diseases (Telfer, 2008, p. 117). Research developments in cancer medicine were presented with little mention of how these might benefit the youngest patients and no published calls for public support for cancer charities mentioned paediatric malignancies. However, children's cancer entered the British press in the 1960s, as British cancer charities adapted the style of their American cousins by personalising campaigns around specific patients and particular researchers, and as research into children's cancers started to demonstrate the improvements in outcome that had long been promised in the American news.

The Foundation of the Leukaemia Research Fund

In 1960, Susan Eastwood died of acute lymphoblastic leukaemia at the age of six. Her parents David and Hilda were heartbroken, but having read a feature in the *Sunday People* newspaper about research into children's diseases, appealing for a new building at the Institute of Child Health (the medical school attached to Great Ormond Street Hospital), they decided to write to the Institute's director, Professor Alan Moncrieff, offering to raise money for leukaemia research. With his encouragement, they established a fundraising campaign in their home town of Middlesbrough, in the north of England, an enterprise that drew support from other families of leukaemia sufferers, local politicians, and the town newspaper. By October, they had set up the Teesside Leukaemia Fund, with David Eastwood as its secretary (Piller, 1994, p. 6).

The local charity raised over £3,000 in its first year of operation, and discussions began at the Institute over how this could best be utilised to support research into leukaemia. Moncrieff met with Dr Gordon Piller, the Chief Executive Officer of Great Ormond Street, and haematologist Roger Hardisty to discuss whether the Institute might be well placed to launch a leukaemia research unit. Hardisty agreed to serve as the honorary head of the new unit, and Piller as director of the charity that funded the unit's work. The Teesside charity continued collecting donations until they reached the £5,000 mark, sufficient to cover a year's work, so the new unit – the first of its kind in Britain – could open (ibid, pp. 9–10). A press release concerning the opening of the research unit was widely carried in national newspapers, including a piece in *The Times* on 7 December 1961 welcoming the opportunity provided by the charity – newly christened the Leukaemia Research Fund (LRF) – to expand the hospital's long-standing interest in tackling leukaemia in children (*The Times*, 1961a).[2] Hardisty's first research programme explored techniques for growing leukaemia cells in tissue culture to enable drug testing, but no promises or optimistic claims were made by journalists about immediate changes in available treatments, or in likely outcomes for affected children.

These news stories did, however, set in motion a flood of public enquiries to the Medical Research Council (MRC) from relatives of children with the disease and from members of the public wishing to raise funds. Relatives of sick children wanted to know how to contact the LRF; whether the LRF's unit was developing any treatments not available elsewhere; and why a charity was paying for this research rather than the British government. Members of the public wrote to ask where their collections of money for leukaemia research should be sent.

The LRF's habit of writing frequent pieces for the press did not meet with approval at the MRC; the two funding bodies' divergent views as to how research grants and progress should be communicated meant that relations between them remained strained over the next decade. The LRF actively courted the news media, which was markedly different from the more private approach adopted by the MRC and most of its grant holders, and there was no tradition of cooperation between the LRF and the MRC, no network of friendships and professional histories between personnel other than Hardisty's appointment, and no history of negotiated collaboration. As we saw in Chapter 2, the MRC had been through a process of careful negotiation with the two major cancer charities, the Imperial Cancer Research Fund (ICRF) and the British Empire Cancer Campaign (BECC), over how resources from different sources could best be deployed in support of laboratory researchers and hospital doctors. By the 1950s, effective working agreements had been reached over which body should be approached for which kind of research, and the three bodies maintained the courtesy of keeping one another informed over their complementary priorities (Austoker, 1988; Edwards, 2007). The arrival on the scene of a new cancer research charity posed new challenges to the network of British funding bodies supporting research in the scientific and clinical realms.[3]

The fledgling LRF made contact with the Charity Commission to ensure it would not be competing with any other charity interested specifically in leukaemia, and its leaders held talks with staff at the ICRF and the BECC to ask if there were objections to the charity's foundation. A medical advisory panel was set up in 1961 to try to ensure scientific rigour in the LRF's research programme, and to secure scientific cooperation with the ICRF and the BECC; the panel included John Dacie and Frank Hayhoe, two of the most respected and prominent research haematologists of the day, who had both supported the MRC's leukaemia conference in 1957 (Piller, 1994, pp. 12–13).[4]

Being newer and smaller than the ICRF and BECC, the LRF needed press coverage in order to expand from its two initial branches in

Teesside and Great Ormond Street into a nationwide body. Staff at the LRF sought as much publicity as possible to showcase the work they already funded and to attract additional donations and coordinators who might establish new, active branches. The MRC's internal staff, by contrast, adopted a much more cautious attitude to making press releases, on the assumption it was better not to announce anything relating to leukaemia or tumours in children until the Council had definitive news to share.

The LRF not only promoted its research more often than other bodies felt appropriate, but it also did so in a more emotive manner, consciously

LEUKAEMIA
NEWSLETTER

Leukaemia Research Fund
Registered National Charity supported
entirely by voluntary donations

No. 12   April 1972

THIS IS WHAT THE FUND IS ALL ABOUT

This dramatic story reprinted from "The Sunday People" newspaper sums up what the Leukaemia Research Fund is all about. For Jason and all the hundreds of other children stricken with leukaemia we are resolved to work unceasingly until this tragic disease is brought under control. But there is no generation gap in leukaemia — it affects adults too in even greater numbers.

This year the Leukaemia Research Fund has spent more money to finance research than ever before. Grants for research in the year ending 31st March 1972 total £133,000. This shows how earnest we are to fight leukaemia and to back the doctors to the hilt. Something must give soon as the research programmes expand at many important medical centres in Britain. The Fund is dependent on voluntary donations to support its urgent and vital work. We must intensify our efforts to beat leukaemia. Please help make 1972 a year of greater progress by giving to the Leukaemia Research Fund and supporting your local Branch.

The fight against leukaemia is a battle that must be won.

This little chap needs your help

"He looks so contented in his wonderful world of make-believe.

And if you saw him romping with Simon, his nine-year-old brother, you would never believe there is anything wrong with Jason Stevens.

Often as good as gold. Sometimes a little terror. Had a fight at school just before the end of term. Goodness knows what he has been getting up to over the holiday.

Just a normal, happy, six-year-old scamp?

Alas, no. Jason, who lives in Staines, Middlesex, has leukaemia.

His illness was diagnosed three years ago and all his parents can do now is hope the drug treatment Jason receives constantly will achieve the miracle they pray for. It's a fervent hope, but just a hope.

The next few years are crucial in leukaemia research for at last the doctors are beginning to close in on this terrible disease.

But they need more cash if they are to save smashing kids like Jason. They need YOUR cash — just a bob or two will help.

Reprinted by kind permission of The Sunday People.

Figure 5.1   Leukaemia Research Fund newsletter
Source: Reproduced by permission of Leukaemia and Lymphoma Research.

drawing attention to the human cost of childhood leukaemia in its publications. As we have seen, the LRF was closely aligned with Great Ormond Street Hospital, which had a long tradition of using emotional advertising techniques to solicit donations, and the LRF followed the same practice of seeking high-profile advertisements in the press, and famous people to lead appeals.[5] In November 1964, Moncrieff used his connections within the BBC to get the LRF nominated as a 'Week's Good Cause': the radio appeal raised over £3,000 in a week, close to the record level for charity fundraising in Great Britain at that time. The following year, comedy stars Hattie Jacques and Kenneth Williams supported the production of a colour film to be screened at fundraising meetings in towns and villages, a technique that had successfully built a supporter base when used by the American Cancer Society in the

Figure 5.2   Leukaemia Research Fund poster
Source: Reproduced by permission of Leukaemia and Lymphoma Research.

1940s (Piller, 1994, pp. 18–19). In 1967 the LRF launched a trading arm to sell Christmas cards and gifts, personifying the quest for a cure by using a picture of a crying girl crushed by a tower of the word 'LEUKAEMIA' on the catalogue cover. These emulations of American campaign tactics sustained the charity's continuing growth in branches, donations, and funded projects; this approach was also in stark contrast to the drier tone adopted by the BECC and ICRF in their newspaper advertisements from the same period (Figures 5.1 and 5.2).[6]

Tales of Heartache

Over the course of 1962 and 1963, the MRC received dozens of letters from members of the public asking for more information about the LRF or mistaking the MRC for the LRF. These have been preserved in two files in the National Archives in Kew; they serve as a valuable window into the level of public awareness and understanding of leukaemia in the early 1960s, when chemotherapy trials in the USA and Great Britain were just beginning to deliver a very small but nonetheless significant proportion of cures. To preserve the privacy of those families with children with leukaemia, who had no intention of making their identity known to the public through writing to the MRC – unlike families featured in newspapers or biographies, for example – we have removed the names and some personal details of correspondents.

The first letter received by the MRC was typical of many, the author not being aware of the actual name of the LRF. She wrote in November 1962 requesting the address of the 'Leukaemia Research Council'.[7] Dr Malcolm Godfrey, Second Secretary at the MRC, replied that there was no such body, suggested 'you may be thinking of the Leukaemia Research Fund', and supplied the correct address.[8] Harder for Godfrey and his colleagues to deal with were the letters from distressed relatives, desperate for any additional help or hope there might be for their loved ones. One such came from a bereaved mother about her son:

> Our son ... aged nearly 3, died in Kings College Hospital on Sept 30th from leukaemia, which was only confirmed 3 days before he died. I have since then been trying to get details of the leukaemia research fund as we wish to contribute to it and if possible, keep a collecting box in our home.
>
> We would also be interested to know whether it has been proved that the illness is <u>not</u> caused by radiation? Also, has the disease increased and become more malignant and acute, as our son's death

was so sudden and rapid? Previously, I understood that a person could live several months or even years with leukaemia.

I should be most grateful for any information you can give us as to the fund and also on current research and possible cure.[9]

Godfrey set in motion the process of collating information and drafting a reply to express sympathy and to explain the changing incidence and virulence of the disease. It would appear that this was the first such letter the MRC's staff had needed to write: it went through three drafts over the course of a week and was discussed with a number of colleagues before being signed and sent. It then served as a template for future letters to distressed relatives of leukaemic children.

The bereaved mother had asked whether the apparent increases in the rate and severity of the illness were real and if radiation was a causal factor, forcing the MRC staff to revisit their debates of the 1950s, which, as noted in Chapter 2, had led to the formation of their leukaemia working parties, and to the committee charged with reporting to the Prime Minister on the hazards of radiation. The carefully crafted reply was designed to give truthful answers without provoking undue alarm or hope:

It is, unfortunately, true that the trend in incidence of leukaemia is rising. But it is important to appreciate that such an increase was in evidence before 1945 and, indeed, has in fact been recorded since 1920, so it cannot be entirely attributed to any increased levels of radiation. Ionizing radiations are known to be capable of causing the disease, and it is possible that some of the increase may be due to their more extensive use; however, it seems probable that other factors are also involved. If you wished to follow this up this whole question was dealt with in some detail in Appendix A of the Council's report on 'The Hazards to Man of Nuclear and Allied Radiations' published in December, 1960. You would, I expect, be able to obtain a copy of this on loan through your local Public Library.

As regards the malignancy and rapid course of the illness this is not necessarily due to any recent change in the nature of this disease, since it has been known for many years that the course of the disease may follow either an acute or chronic pattern. In infancy and childhood the disease is very often rapidly progressive, with death occurring within a few months or even a few days. The chronic form of the disease, which is rare in children, progresses more slowly and it may be some years after first diagnosis before such people die.

At present there is, I am afraid, no specific cure for leukaemia, but work is being carried out in several research departments through-out the country, and a leukaemia research unit was opened at the Hospital for Sick Children, Great Ormond Street, at the end of last year.[10]

The need to educate the public about the forms and expected course of leukaemia, and to gently break the news that no curative therapy existed, was always balanced in the MRC's letters with this closing reas-surance that much was being done which might deliver more hopeful prospects in time. The authors repeatedly emphasised that the increase in incidence could not be entirely laid at the door of radiation: 'indeed' and 'in fact' the increase predated the bombs. For David Hewitt, in his 1954 memo discussed in Chapter 2, radiation could still be the culprit for increases dating back to 1920, as medical uses of radiation and industrial nuclear plants had been increasing the amount of radioac-tivity in the environment since that date. The MRC's reply, however, assumed that this mother was thinking solely of radiation from atomic weapons, and responded accordingly.

Over the next six years, the MRC received a steady stream of let-ters from close relatives of children who were living with or had died from leukaemia. Most were brief, telling us little of the children's treatment or of the information about leukaemia that families had already been given by those treating their children. In 1965, letters also began to arrive from trainee nurses and doctors, studying as part of their training the changing natural history of leukaemia and the new medical and emotional demands placed on children and their families by the more aggressive chemotherapy protocols under test across the country. Replies to these requests for information, and to letters from family members asking if any experimental treatments were available beyond the standard National Health Service (NHS) regimes, were always signed by the same member of MRC staff – a secretary, not a clinical researcher – and the replies themselves were standardised and terse.[11] Trainees were informed the MRC had already published everything it knew about leukaemia, in the offi-cial reports of 1956 on the hazards of nuclear power, and 1957 on radiation given as therapy for ankylosing spondylitis. Relatives were told there were no experimental therapies demonstrating promise that were not accessible through every NHS hospital (MRC, 1956; Court Brown and Doll, 1957). The latter statement evidently left some parents unsatisfied, as we shall see in the final section of this

chapter when we follow the case of a child sent abroad for alternative treatment.

Perhaps the most troubling letters the MRC received were those from
members of the public, or even Members of Parliament (MPs), asking
why leukaemia research had to be funded from charitable sources at
all, rather than the full costs being met by the MRC's budget. On 13
March 1963, the Office of the Minister for Science, then Lord Hailsham,
forwarded a copy of a letter received from the mother of a dying girl,
wanting to know why the LRF was so short of cash, given the obvious severity of leukaemia and the clear emotional costs of watching
children die.[12] In highly emotive language she demanded to know
why the Leukaemia Research Unit in London was not funded by the
government:

> ... [it is] ... run entirely on money collected by us Leukaemia moth-
> ers, and fathers. It seems to me so terrible that they should have to
> rely on private money.

She concluded by calling on Hailsham to think of his own new baby:

> find, if you can, some money to help the Leukaemia Research Unit.
> They may be able to keep my ... [daughter] ... alive for a year and in
> that year perhaps they will find the cure. <u>Please</u>.[13]

The MRC told Hailsham they had never been asked to contribute to the
Unit in Great Ormond Street, but were willing to maintain close links
with the LRF.[14] The MRC tended not to fund projects simultaneously
supported by the BECC or the ICRF, to avoid muddling the lines of
credit for, and control of, research.

Hailsham replied to the mother as follows, giving a rationale for the
charitable funding of research:

> how deeply distressed I am about your little girl who happens to
> have the same Christian name as my own. I can imagine what you
> must be suffering at this time ...
>
> A great deal of research is being carried on to discover the causes
> and possible cures of leukaemia and other diseases which have hith-
> erto defied solution. Most of the support given by the Government
> to this end is applied through the Medical Research Council whose
> total budget is about £6 million a year. The Council carry out

research at their own units and also support research at universities. They also make quite big grants to other research centres such as the Institute of Cancer Research, the Strangeways Laboratory, and the Christie Hospital and Holt Radium Institute ...

Financial support for research on leukaemia is provided in addition from independent sources, notably the Imperial Cancer Research Fund, the British Empire Cancer Campaign, and the Leukaemia Research Fund.

I am often asked why the Government does not step up its expenditure on research into illnesses sufficiently to make the work of these voluntary collecting organisations unnecessary. The reason is that no one is infallible and professional opinion differs quite widely on what is and what is not a promising line of research. This means that official organisations do not have, and thus must not pretend to have a monopoly of wisdom. The conclusion is that they must not be the only dispensers of money and there should be independent foundations of voluntary bodies having the freedom and resources to choose alternative lines of research for support and to act as a court of appeal if particular research projects are not approved. These organisations normally work in harmony with the Medical Research Council, and I understand that the Leukaemia Research Fund have expressed a wish to maintain close contact with the Council ... If I thought that any worthwhile scheme for research into Luekaemia [sic] was being held up for want of money, I would immediately ask the Medical Research Council to look into it. But they assure me that this is not so ...

I realise that the eventual prospects of success in which I believe can be little consolation to you. What you do need to know, and what I hope will comfort you is that money is not the limiting factor in this research, but the poverty and fallibility of human ideas. If money were what was lacking I should certainly find means of getting it.[15]

Whether or not members of the public, members of parliament, or parents of sick children believed this concluding argument, it was one made repeatedly by the British government when asked to defend its constrained budget for medical research.[16]

Certainly the research scientists themselves did not agree. Professor Sir Alexander Haddow had written to *The Times* two years earlier, dismissing the claim made in the House of Commons by the then Parliamentary Secretary for Science, Mr Denzil Freeth, that it was 'a relative lack of new ideas' and not money that was holding back the quest for cure:[17]

> With respect ... this statement is over-simplified, misleading and complacent ... Nothing could be further from the true state of affairs. Over the past 15 years, the field of cancer research has been amongst the most active in the whole of biological science ... [I]deas have proliferated at a rate almost embarrassing from the point of view of those who have to study and test them from day to day (Haddow, 1961).

The funds available from British sources – the two big cancer charities and the MRC – could not cover the costs of even the most promising projects, the shortfall having to be met by the American National Cancer Institute (NCI).

Haddow's general argument rested on clear facts: there were always researchers calling for funding over and above that available from the most prestigious British sources. Indeed, the internationally famous Oxford Survey of Childhood Cancers depended on American funding for much of its lifetime.[18] But the government line that there was a shortage of *viable* ideas was also easy to justify. The LRF, for instance, by the late 1960s, had more funds than it knew what to do with. Public generosity swelled the coffers of British charities to the point that staff wrote to one another asking if there were any 'good' proposals going spare. The LRF endowed a Chair in Haematology at Cambridge University in 1968, taken up by Frank Hayhoe who researched the typing of leukaemia and identified rare forms. They also funded an international symposium on leukaemia research in humans and animals, and initiated an annual guest lecture with resources to cover international air travel for the speaker and publication costs, yet they still had spare funds.[19] Piller, as Honorary Secretary to the LRF, consulted with MRC staff over the possibility of working together in order that the MRC could pass on good projects that lay outside its remit or budget.[20] For leukaemia at least, it was possible for the government to argue there were more funds than ideas. This was part of the challenge for the MRC: while public funds might flow freely to alleviate children's suffering, the priorities of private donors did not match up precisely with the research priorities of the medical science elite.

In 1960, cancer research had belonged to the MRC, the ICRF, and the BECC. These three bodies had worked very carefully over the preceding 40 years to develop good working relationships. By the time the LRF entered the world of cancer campaigning, these three had traditions of sharing personnel, and comparing proposals, to ensure that no organisation applying for money obtained funds from multiple sources without all funders being aware of the fact, and that no researcher deemed a 'safe bet' would be turned down by all three. Cancer research institutions were encouraged to apply to the body most interested in their line of work, and if individuals contacted an inappropriate funder, then staff within the three bodies would redirect applications, as we saw in the case of research concerning Burkitt's lymphoma. The LRF, with its base at the Great Ormond Street Hospital, *had* to be factored in, because of its location, and because Roger Hardisty, the steering clinician for the MRC's leukaemia trials programme, worked there. Cooperation between the LRF-funded unit and the MRC's clinical researchers had to be established to avoid disruptions to the funding structures for biomedical research into cancer.

The Competing Attractions of Cooperation and Independence Between Funding Bodies

Much of the newspaper coverage of cancer through the early 1960s derived from annual statements of the funding bodies. The public was educated about the total cost of cancer research and the spread of work across the country, but rarely did these accounts explain how grants were spread across projects, or reveal the real cost of any particular line of research. In 1960–61, the MRC announced it had spent £650,000 on cancer, but this was put into the shade by the BECC, which made grants totalling more than £1 million to support basic work on cancer biology in units all over the British Commonwealth, including Uganda (*The Times*, 1961c).

For every new therapy and discovery surrounding cancer's aetiology, the funder gained publicity – crucial for charities trying to maintain good subscriptions, but also for a government seeking to present itself as properly addressing the public's concern with cancer. The ICRF was given positive coverage after its 60th annual report in April 1963, when the research director, Dr G. F. Marrian, drew attention to the Fund's support for a group in Entebbe exploring the possible viral aetiology of Burkitt's tumour, and the connections between this work and cutting-edge research being conducted in laboratories in Great Britain and the USA on virus-like particles detected in various other human and animal cancers (*The Times*, 1963a).[21] Favourable reports were also made concerning the MRC's funding decisions when the Council demonstrated that government-funded work on the carcinogenic properties of human adenoviruses had borne fruit.[22] In the early 1960s, any body funding work into cancer viruses could secure glowing reports and generous amounts of page space.

The MRC, ICRF, and BECC had learned to work together well, but the arrangement had its critics, and not just from the LRF. In November 1964, *The Times* (1964d) published a long, anonymously written article outlining the relative strengths of the various funding sources, their different *modi operandi* and interests, and the gaps left unfunded. The article, entitled 'A Cancer Research Question Mark', peered behind the seemingly happy relationship between state and private funding, and made a radical call for merger. To be sure, cancer research was prominent in the news, but perhaps the diversity of reports, on a plethora of putative causes and possible therapies, indicated it was time to revise the battle strategy:

> for all the brilliance, of the brains involved, for all the years of effort and the millions spent, the problem of cancer remains unsolved ... which raises the questions of whether cancer research in this country is being approached in the right way, whether it is being organized efficiently, and whether enough money is being made available.

The two dominant British charities were strong and the article recounted their good points, emphasising how they complemented each other: while the ICRF could undertake large-scale capital building projects, the BECC could respond quickly to research leads with small grants.[23] Good communication between the charities, and with the MRC, ensured there was no 'unnecessary duplication of research' and that every promising lead found a source of support. Nonetheless, the writer felt the situation could be improved:

> Would it benefit the cause of cancer research if the two major charities were to merge, or if some means were found of drawing them together with the Medical Research Council ... ?[24]

The three bodies opposed merger, which they felt would make the awarding of grants too bureaucratic. But centralisation of the awarding of research monies would make it far easier to ensure that insights crossed from laboratory to laboratory, and from discipline to discipline, in a more organised and timely fashion, speeding the pace of progress.

The article closed by suggesting it was not money *or* ideas that were in short supply, but cooperation:

> the forces of tradition and the desire for independence are strong ... [but charities should agree to] ... some form of rationalization in the structure of the research organization which could bring more intimate cooperation between the many research centres (ibid, p. 15).

This argument for rationalisation had merit – and indeed the two charities did merge in 2002. But diversity of funding sources also had its advantages, as Lord Hailsham had noted.

The existence of nongovernmental funding sources demonstrably gave researchers greater freedom to publish results that went against the orthodox view of cancer aetiology and treatment. For example, in the case of Alice Stewart's work on antenatal X-ray exposure, the MRC was very concerned when she communicated her preliminary findings – from the initial returns to her survey of childhood cancer, published in *The Lancet* in 1956 – without seeking 'proper' authorisation from the MRC working party of which she was a member. But as MRC papers indicate, as she had been funded independently, its staff felt it had little right to restrict her publications. Joan Faulkner wrote to Dr Williams of the MRC internal staff concerning the proposed publication, fearing Stewart had not gathered enough scientific data to warrant making the controversial claim that standard obstetric practice was harming the unborn:

> Since she started the work before our Leukaemia Working Party was set up I doubt we are in a position to instruct her to delay publication ... On the other hand she was given the grant by the Lady Tata Memorial Fund on condition that the work was carried out under the supervision of the Working Party. This may have been Dr Green's way of keeping control over her notorious enthusiasm![25]

Given the role of charitable funding in the research, the MRC could only request that Stewart checked her argument with Sir Bradford Hill prior to publication, not insist she delay.[26]

The MRC strived to preserve the diversity and independence of research supported by private funds. It kept its distance from all private funding, and preserved its claims on government, by declining financial gifts from charities with more cash than fundable proposals. Even in the late 1960s, when the MRC was financially pressed, Principal Medical

Officer Dr Brandon Lush wrote to Dr Julie Neale, who handled the correspondence with Britain's leukaemia researchers:

> [it is] ... highly desirable that there should be alternative private sources of support for medical research – particularly at a time when public funds are short – but equally it would defeat the purpose of private funds if they were simply used to relieve public funds.[27]

Independence of resource allocation was not incompatible with communication about who funded what, but such communication was certainly lacking between the MRC and the LRF. The LRF had rapidly acquired a reputation for talking to the press without giving the MRC prior warning, even while asking to be 'kept in the loop' about all the work funded, and turned down, by the MRC.[28] We saw in Chapter 4 how carefully negotiations were carried out over who should, could, and would fund Burkitt's work. The LRF wished to be party to such discussions and have similar working arrangements with the MRC, but MRC staff remained reluctant, sensing the LRF did not operate in the same way as the MRC itself, the BECC, and the ICRF. Perhaps the MRC simply feared change and thus did not offer the LRF sufficient guidance on ways of cooperative working which the MRC would find acceptable; but it appears there were real differences of opinion about how leukaemia research should be conducted and made public, and that the LRF senior staff would not give way and accede to the MRC's line.

Dr Gordon Piller served the charity for 30 years. His views on leukaemia research and his management style, however, were anathema to the MRC staff, which typically resisted being drawn into public debates about cancer research funding and the possibility of future breakthroughs. Piller, in contrast, frequently gave talks and wrote articles about leukaemia and how best to advance its cure. He addressed the general public through newspapers, and issued press releases about research teams being funded by the charity; both tactics were problematic for the MRC and the Ministry of Health.

In the spring of 1964, an article by Piller on the state of leukaemia research was published in at least two local newspapers, possibly more. *The Newcastle Journal* published 'Leukaemia's Challenge' on 24 March and *The Sheffield Telegraph* printed a slightly longer version as 'The Whole World Waits for a Breakthrough' two days later.[29] The latter had an additional introductory paragraph:

Leukaemia victims – and particularly the parents of child sufferers – are still hoping desperately that the Naessens 'cure' will indeed prove to be the answer to this dread disease. In the meantime constant and exhaustive research is being continued on a wide front.

Piller went on to outline the places where research was being done, and what had so far been gathered about the nature of the disease. While chemicals could provide temporary relief from the disease's progress there was, as yet, no known form of treatment to which the disease did not become resistant. Piller had little faith in the 'Naessens "cure"' – which we will explore in more detail shortly:

A lot has been heard of leukaemia in recent months. Claims of cures are reported from time to time from all parts of the world. These produce hope, but seldom add anything to what is already known of this disease.

He closed the article by suggesting that a sturdier form of hope, a grounded hope, lay instead in supporting conventional research into the disease:

a great deal of research is going on and will go on until a cure is found. In the last 20 years our knowledge has considerably advanced and everyone is now waiting for a breakthrough.

This bold optimism was not to the MRC's liking. Such statements had in the past drawn questions in the House of Commons and led to an influx of letters from relatives of the sick about exactly when cure would come. Piller's tendency to promote his charity and its generosity also frustrated MRC staff. In the 1964 articles, Piller – naturally – described the work of the LRF's own unit at Great Ormond Street first, but in a manner which made the MRC's contribution to the field appear less significant:

The Leukaemia Research Fund, a national charity for research into the disease, is supporting research at Great Ormond Street Children's Hospital, Derbyshire Royal Infirmary and at Newcastle University.

The Medical Research Council is also supporting research of various kinds as are some other charitable bodies.[30]

This dispute over credit for funding specific projects came to a head a decade later when Piller issued a press release concerning the LRF's support for a group researching immunology at the Institute of Cancer Research.[31] The release stated that the LRF had distributed three quarters of a million pounds since its initial grants in 1962. As we have seen, the MRC was awarding close to this amount for cancer research every year. What riled MRC staff even more was that they were already funding the same research team. There had been no consultation or warning, and thus the MRC were, so to speak, dropped in it, when called upon by journalists to comment on the LRF story.[32] Piller was asked for more caution and cooperation in future dealings with the press, given the by then huge media interest in any news regarding leukaemia research and possible treatment breakthroughs.[33]

MRC press and correspondence archive files from the 1960s are peppered with statements that reiterate a desire to avoid making comment when cancer cases were discussed in the news, the Houses of Parliament, or the correspondence in-tray. Whether it was children being flown overseas for treatment, servicemen claiming war pensions for exposure to nuclear tests, or employees filing claims against their employers for damage supposedly caused by exposure to carcinogenic chemicals or ionising radiation, the MRC – and the Ministry of Health – tried to say as little as possible in public.[34] Piller and the LRF were not likely to agree to follow the same approach: the charity depended on individuals' donations, with its keenest supporters having lost relatives to leukaemia, and strongly wished to keep the disease in the public eye and on the government agenda.

In the spring of 1964, shortly after Piller's articles referring to Naessens, a team at the NCI's clinical centre in Bethesda announced it had developed a multidrug regime for acute leukaemia that appeared capable of killing every leukaemic cell in a small proportion of its human patients: treatment that could deliver cure as opposed to remission. This breakthrough did not make front-page news in the mainstream papers, but was given coverage in the general press, as well as considerable page space in *The Lancet*. The knowledge that chemotherapy did have the potential to cure acute leukaemia meant orthodox medicine could, at last, provide hope. In the next chapter, we will describe the development and expansion of curative protocols as they grew more effective and reached higher proportions of children with an expanding range of malignancies. But here we look first at press coverage of Naessens' 'breakthrough' from a few months earlier. It proved a false hope, but the coverage underlines the continued hopelessness of conventional approaches and the recurrent attraction of foreign remedies.

'Wild Hopes in Desperate Hearts': Anablast and British Leukaemic Children in the News

In 1963, the national press covered the story of one boy from Blackpool and his family's efforts to get him treated by a French 'biologist', pushing the plight of children with leukaemia into the spotlight. The 'biologist' in question, Gaston Naessens, had no recognised diploma in science, leading French authorities to dispute his credentials. Nevertheless, he featured frequently in articles both in broadsheets and tabloid newspapers over the next four months. His treatment for cancer was known as 'Anablast' – literally, without cancer cells; it was said to involve a serum derived from horses into which cancer cells had been injected to produce antibodies.

The boy, Edward Burke, four years old and the youngest of three children, had the support of his Councillor, Raymond Jacobs, and the townspeople of Blackpool, who raised £4,000 to send him for injections of Naessens' preparation. He also secured mass support from the people of Corsica, and then of British practitioners and promoters of other alternative remedies for cancer. His Councillor and parents wrote to the MRC, to Members of Parliament, even to the Queen, seeking further support. Their letters have been preserved as a matter of public record in the National Archives, allowing us to build up a multifaceted account of this story, with access to many viewpoints as to what was the best thing to do after conventional treatment had failed.[35]

It was on 22 December 1963 that Edward Burke flew to Corsica to be treated in the regional centre, Ajaccio. Naessens was in the semi-independent Corsica because no official laboratory had tested his anti-cancer serum, and the French authorities would not allow Naessens or any French doctor to administer it to patients. But even in Corsica no doctor could initially be found who would inject the boy with Naessens' preparation. The British Vice-Consul in Corsica, Mr Snook, who had arranged for the child to fly out with his mother, announced on 27 December that unless one was found by the following day, the pair would have to return to Great Britain. That evening, 3,000 Corsicans staged a rally, demonstrating their sympathy for the plight of the boy with slogans reading 'Quickly, leukaemia does not wait' and 'We must save Eddie'. The next day, a prominent local doctor, Dr Santonacci, stepped forward and Edward began treatment two days later.[36]

The French authorities grew alarmed at the scale of publicity Naessens was attracting, and issued a statement that the serum was not to be given to any patients in Corsica, the penalty for any doctor breaking the prohibition being a revocation of his or her licence to practise. In Britain, the Ministry's staff realised they needed to plan for an influx of enquiries from families in similar situations, and make arrangements to manage the upset resulting from the French ban on Anablast's use. The Blackpool medical authorities warned the Ministry that requests for treatment could well 'snowball' and suggested that – in order to prevent a wave of desperate parents taking children overseas – the serum's efficacy should be ascertained as soon as possible.[37]

The news of the Burke family's voyage stimulated many other families to write to Mr Snook, to their Councillor or Member of Parliament, or to the Ministry of Health, attempting to gain access to the same treatment for their own children. Realising the situation could get out of hand, the Ministry, usually determinedly silent on claimed cancer 'cures', had to act. One newspaper noted that more than 100 British families had made requests for their children to be treated in Corsica with Anablast, while Naessens was issuing press statements calling for no more patients in order to conserve stocks of serum for those already being treated (Smyth, 1964). The gaze of the public at large had fallen on leukaemia, and the number of children affected became common knowledge: the *New Scientist* noted the Naessens affair had made everybody aware of the 'peculiarly tragic illness' of acute leukaemia, which almost exclusively and rapidly struck down the young (Gould, 1964a, p. 402).

Letters came from across the country throughout January, requesting the Ministry make the serum available in Britain. A young boy from Glasgow was flown to Corsica to be treated alongside Edward and five French children, and many other families were desperate to gather the funds required.[38] The family of a girl from Orkney asked their MP to seek advice from the Ministry; the parents of a girl in Stevenage persuaded the Public Health Committee of the local council to write to the Minister of Health, Anthony Barber; and a boy from Liverpool was in danger of being withdrawn from the care of Dr Keenan at the Alder Hey Children's Hospital in order to start Anablast.[39]

On 24 January, Barber received a personal letter from Charles John Robert Manners, the tenth Duke of Rutland. His son, Robert George, was ill with a disease of the bone marrow related to leukaemia, and Rutland was wondering whether to press for supplies of Naessens' serum.[40] Rutland was seeking hope wherever it might be found:

I think it is a very great mistake to be too sceptical and over conservative about a matter such as this. After all, some of the great medical
discoveries have been found partly by chance and often by quite
unknown doctors.

Rutland offered to fund tests of the serum's utility, as he felt 'it would be
in the public interest'.[41] He was not alone in calling for British involvement in the verification, or discreditation, of Naessens' claims.

Those close to the Minister agreed the situation needed to be managed in order to prevent scores of British families flying to Corsica where
there was no guarantee that Naessens would treat their children, let
alone that the serum would have any positive effect. Ministry staff spent
January 1964 striving to develop some scientifically grounded basis for
an official government response to the situation. Patrick Jenkins, prospective Conservative candidate for Wanstead and Woodford, wrote to
Barber, his friend, on 30 December:[42]

I have received from a prospective constituent (if I may so describe
him) a pathetic appeal for help to secure treatment for his small son
who is apparently dying of leukaemia. All I have been able to do is
to give him the address of the British Vice-Consul in Ajaccio who
appears to have been instrumental in securing the treatment for
Edward Burke.

At the same time, I am sure you will appreciate how news of
this sort, perhaps quite unreasonably, arouses desperate hope in
the minds of those whose children are afflicted with this disease.
While I realise that proper investigations into a new drug are always
lengthy and complex, I wonder if it is quite impossible for some sort
of preliminary opinion on the validity of the claims to be expressed
officially fairly soon? I have visions of numbers of parents spending
hundreds of pounds on what might be no more than a quack cure,
or alternatively eating their hearts out in misery because they cannot
raise the cash to even try it.

On 6 January 1964, Naessens offered his serum to any government
willing to test it. The Institut Gustav-Roussy in Paris took up the
offer.[43] The laboratory spokespeople warned that full test results
would not be available for up to two years, and from many quarters pressure was put on the Ministry of Health to test the serum in
Great Britain. Jenkins asked Barber if the Chester Beatty Institute
might be able to test the serum,[44] and Manchester's Christie Hospital

volunteered to carry out the assessment (*The Times*, 1964e). But it was a group in Edinburgh, funded by a Scottish businessman, which eventually conducted the first British examination of Naessens' serum and methods. A medical panel was assembled, headed by the highly respected Edinburgh radiotherapist, Professor McWhirter, for two days of confidential meetings away from the eyes of the press, in the middle of January.[45] The panel recommended that the French trials would be sufficient, that the claimed clinical results should be backed up by a controlled study, and that Naessens must immediately arrange for full publication of his methods, observations, and theories. Preparations were made to relocate Naessens to Britain, at least temporarily, thereby ensuring a steadier supply of Anablast for British families and coincidentally keeping Naessens away from French media attention while the authorities in his homeland investigated whether there was sufficient material to bring a case against him for practising medicine without a licence.[46]

Despite the warnings of the testing body, initial results from Paris were made public within a few weeks: scientists at France's most prestigious cancer institute declared the serum to be useless, and gently mocked Naessens' biological theories and skills. Professor Pierre Denoix reported to the French Minister of Public Health that Naessens' work was riddled with errors. The microorganisms that he claimed to have seen present in the blood of leukaemics were, in fact, not new, but had been described and photographed by several blood specialists before him: they were known as myelinic figures, a product of the fragmentation of red blood cells, and – far from being a sign of disease – they were present in the blood of everyone. Microorganisms Naessens had managed to culture from patients' blood were shown to be the product of contamination of samples. He was therefore deemed inept both as an observer and an experimenter. All the people Naessens claimed as 'cured' by Anablast were discovered to have already been treated in the conventional manner; Denoix thus concluded that the results observed with the serum were simply the delayed positive effects of earlier therapy. Any improvements witnessed in the appetite and spirits of patients were dismissed as placebo effects. Denoix declared that Naessens was peddling nonsense: a product that could not possibly have any effect on leukaemia or solid cancers.[47] The French authorities thus declared that no doctor was to administer the serum, including those in Corsica.

The Ministry of Health accepted that to forbid treatment with Naessens' serum would be to court public outcry and that in a sense the British government could not win the public relations war. When

Naessens and the Burke family first sought permission for entry into Great Britain to continue the treatment in Blackpool, a Ministry employee wrote 'we are vulnerable either way, whether we refuse or whether we allow this'.[48] Letters were received from Edward's family, and members of the public, pleading that the last hope of families with children with leukaemia not be taken from them. One of these writers was a supplier of 'Medicated Bee Venom Therapy':

> [I am] ... as a member of the public, as a human being, and as a woman, appalled at the prospect of these little children being denied their last hope.[49]

The Daily Telegraph dismissed the early results from Paris, on grounds that the tests scheduled to last 12 months had been shelved in order that Anablast could 'summarily [be] pronounced valueless' by the French government.[50] Edward's father telegrammed the Minister after the French decision to force the end of treatment:

> I implore you to intervene to save my son ... treatment must not be discontinued at this vital stage otherwise all hope gone.[51]

Jenkins' constituent-to-be broke down and cried when she read of Denoix's verdict on Anablast; Jenkins wrote to Barber:

> It does seem wicked that an absolutely unqualified man like Naessens should have aroused such wild hopes in desperate hearts.

Jenkins closed by dismissing the serum as a 'quack remedy'.[52] The Ministry had to find a line to tread between being seen to close off hope and appearing to promote an unproven treatment over the chemotherapy regimes then available in the larger children's hospitals.

Barber's staff quickly developed a statement on which the Ministry could draw when asked for advice in specific cases. His memo of 28 January stated that the Foreign Office had been told to allow small quantities of the serum to be imported without hindrance for use in particular patients, without the need for a licence as was usually required under the Therapeutic Substances Act. Further, Barber made it clear that doctors would not be penalised if they chose to administer Anablast: the matter was to be left to the judgment of individuals.[53]

Appeals from Councillor Jacobs, and Edward's mother Mary Burke, that the Minister attempt to prevent the French Ministry of Public

Health from prosecuting Naessens, were unsuccessful. Mary Burke telegrammed Queen Elizabeth, appealing to their common motherhood and requesting her help in protecting Naessens from governmental attacks.[54] Other families struggling to cope with the slim chances for children with leukaemia also kept up the pressure during February, as did Naessens himself, appointing a lawyer to explore the legality of producing the serum in Britain.[55] Meanwhile, the Ministry continued to handle letters from the parents of sick children, or their elected representatives, and each time the advice remained the same: approach your local hospital as all treatments which have shown clinical promise are available through the NHS.[56] Staff in the Ministry strived to be as unhelpful as possible to Naessens, while stopping short of blocking the use of Anablast. The stated aim was to keep him out of the country to avoid further falsely raising the hopes of families and potentially seeing children withdrawn from orthodox treatment to be treated by the worthless serum.[57]

By April 1964, the story had largely disappeared from the British press. A new story about leukaemia was making the news: *The Daily Express* covered the exciting results coming out of the USA, where high doses of four drugs administered in combination had apparently cured four out of 17 children given the experimental cocktail.[58] At last, the Ministry had proof that orthodox treatment for leukaemia, which involved administering steroids, antimetabolites, and other chemicals with proven anticancer effects, could do more than simply slow the disease's progress. The medical profession now had working answers to leukaemia, giving hope to families of leukaemic children: peddlers of alternative remedies would be easier to dismiss.

Naessens returned to the headlines in May the following year, when it was reported that a French court had fined him for fraudulent medical practice.[59] By this date, he had emigrated to Canada, where he established a new laboratory, and began developing a product christened 714-X, derived from camphor and promoted as a treatment for cancer and 'other immunologically based diseases'.[60] The date of Edward Burke's death is not recorded in the Ministry's files – when all hope of cure had gone, the boy disappeared from the newspapers and from official attention.

Providers of alternative medicine, however, continued to attract attention and customers, including families of children with cancer, and further press battles took place between proponents of complementary and alternative medicine (CAM) and orthodox cancer experts. We close this chapter by considering the wider cultural context, to identify why such therapies were attractive to cancer sufferers and the concerned public alike.

Alternative Answers to an 'Unnatural' Disease

The 1960s witnessed the beginnings of a widespread disillusionment with mainstream biomedicine, despite a long catalogue of medical triumphs against infections and chronic illnesses over the previous century. Awareness of the severe side-effects that could result from medical treatment was increasing, with events such as the thalidomide debacle providing particularly graphic evidence that modern medicine was not straightforwardly beneficial for all patients. Many people, if not already actively looking for alternatives, were open to suggestion (Coward, 1989; Whorton, 2002). Unorthodox practitioners, who tended to take a holistic approach to health and offered treatments that appeared more 'natural', found receptive audiences seeking relief from 'artificial' mainstream medicine, deemed by many to be unduly technological, invasive, and dehumanising (Seager, 2003; Klawiter, 2004). Practitioners of long-established alternative methods, including homeopaths, chiropractors, osteopaths, and acupuncturists, were ready to accept new clients. These patients tended to have chronic conditions, including asthma, arthritis and rheumatism, back pain, skin disorders, multiple sclerosis, and cancer, as well as psychological conditions, such as depression or anxiety. What these disorders typically have in common is a tendency not to respond swiftly or straightforwardly to biomedical intervention and to run an idiosyncratic and unpredictable course. Some patients, frustrated by short appointments and ineffectual or unpalatable conventional treatments, chose instead the extended, individualised consultations and ongoing investigations offered by alternative practitioners, often mixing and switching between modes of treatment so as to maintain hope long after conventional options had been exhausted (Furnham and Smith, 1988; Thomas et al., 1991; Sharma, 1992; Murray and Shepherd, 1993; Zollman and Vickers, 1999).

Historians of cancer have studied CAM beliefs and practices, particularly how they have shaped cancer meanings and treatments for adults; CAM is less well studied in relation to children, where the sense that cancer represents an unwarranted disaster and acute tragedy may be greater.[61] Despite the prominence in media history of a few desperate families, it seems only a small proportion of parents sought alternative therapies. This may be because parents seem more likely to follow medical advice for their offspring than for themselves and because national governments retain the right to overrule parents who refuse to consent to medically recommended treatments.[62] There is also an issue about the presumed causes of cancer, which CAM enthusiasts typically frame as an 'unnatural' disease, the result of emotional, nutritional, or environmental toxicity. As the opportunities for exposure to severe damage are few in the first years of life, malignancy in childhood may be hard to explain in this way, even if we grant that children may in some ways be more 'susceptible' to toxic influences.

Nonetheless, some families of child cancer patients in the 1960s and 1970s did turn to CAM therapies, which offered families a degree of agency in selecting treatment options, and supplied hope when conventional treatment often did not. Whatever the presumed cause of cancer, the threat to young life was acute, felt by friends and neighbours, as well as parents, hence the publicity and official concern. In the 1960s, three-quarters of all children diagnosed with cancer died from the disease within five years; this translates to more than 1,000 children annually who could not be cured through orthodox cancer treatment, a large potential market for alternative visions of cancer and treatment (Stiller, 1994). When a cancer patient was faced with an apparently hopeless prognosis, well-meaning friends and family, desperate to help, were likely to bombard the patient or his or her family with any information they could find (Holland, 1982).

The MRC's files preserved in the National Archives contain a small number of letters from parents, neighbours, and doctors seeking advice on whether to try alternative therapies against childhood cancer. In addition to Naessens' serum, bee royal jelly was seen as a possible anticancer treatment in the 1960s.[63] The MRC again coordinated the creation and maintenance of an official government line: that the alternative therapy had not been shown to be effective, and should not be selected in preference to conventional treatment. They also voiced the concerns that, as with Anablast, families might spend monies they could ill afford on an unproven product. CAM therapies were almost invariably only available privately, not covered by medical insurance schemes in the USA or by the NHS in Great Britain.[64] Certainly, the cost of seeking alternative treatments concerned the MRC secretarial team charged with replying to families requesting information about alternative preparations.

After the creation of the LRF, a new worry was expressed: that communities would raise sums to pay for ineffective alternative treatments, instead of channelling funds to research that stood some chance of delivering results. One such case reached the attention of the MRC in

October 1961, when a Lancashire general practitioner (GP) asked for assistance in managing the neighbours of a boy he was treating for leukaemia, then responding well to treatment with steroids and blood transfusions:

> The parents are having a disturbing time from neighbours, who keep bringing them in newspaper cuttings about the Royal Bee jelly treatment, and I have been placed in a difficult position.
>
> I have been approached by the Vicar of the Church for advice about raising funds to bring Dr. Gautrelet over to see the boy, – and I understand, in fact, that a fund has already been started.
>
> I did not feel competent to advise on this, as one could easily encourage a cruel ramp, – but I did not discourage it, as the fund could be useful if directed towards the Leukaemia Research Fund.

Malcolm Godfrey of the MRC consulted experts at the Chester Beatty Institute for their knowledge about the use of royal jelly against cancer. The Institute replied that there was 'no satisfactory evidence that the substance is of any real value in the treatment of leukaemia or malignant disease', supplying a list of references to tests conducted in 1959 and 1960 showing the jelly to be ineffective, and noting that 'Professor Haddow has found royal jelly inactive against his experimental tumours'. Godfrey replied to the GP accordingly, and sympathised with his 'difficult problem'.[65] But, ultimately, the MRC had no power to prevent parents seeking alternative treatments for their seriously sick children.

It is difficult to estimate the proportion of families of children with cancer that accessed CAM providers instead of, alongside, or at the end of conventional cancer treatment, as, by their nature, the therapies recommended by practitioners were offered outside the hospital system (and outside official records), in hired consulting rooms or private homes. The MRC files, however, demonstrate that when cure was not to be found in a cancer ward, some sought far and wide for alternative routes to wellness.

Upholding Orthodoxy

Between 1960 and 1990, the survival rates for all childhood cancers increased incrementally, until more than half of children were living for at least five years after their diagnosis (Stiller, 1994). In Great Britain, those detecting and treating childhood cancer grew more organised in their efforts to standardise experimental protocols and to increase the rates of referrals to specialists in order to maximise the potential benefits of research, for science and for sufferers. As conventional treatment began to offer more success, measures to limit the risk of families abandoning orthodox treatment in favour of alternative remedies were introduced, including providing information about ways in which some alternative treatments might safely be used alongside conventional protocols and restricting the network of institutions that had access to chemotherapeutic agents to those who would follow shared protocols. In the next chapter we describe the centralisation and professionalisation of paediatric oncology in Great Britain, and examine how the treatment options presented to families were modified as a result.

6
A New Breed of Doctor

In the USA, the big push against acute leukaemia in the 1950s, led by the National Cancer Institute (NCI), had led to network of interconnected centres, experimentally testing models of the disease and trialling drug regimes in a programme that, by the early 1960s, included patients who had survived over five years since diagnosis. As we saw in the last chapter, with the major breakthrough using combination therapy in 1964, acute leukaemia began to look curable for some children (Freireich et al., 1964). A number of these cooperative groups launched chemotherapy trials for children with solid tumours alongside those for leukaemia, and published reports of incremental or dramatic increases in survival lengths and rates of cure for many of the most virulent cancers affecting children through the 1960s.[1] By the end of that decade, researchers and clinicians could state with some confidence that childhood cancer research in the USA was well organised and delivering improved outcomes for patients suffering many of the paediatric malignancies.

Clinicians specialising in the emerging field of paediatric oncology in Great Britain also grew more organised in the 1960s, joining American and European professional bodies to keep abreast of the latest scientific disease models and experimental therapies, and ultimately forging a British professional association for the discipline. As a result of their lobbying efforts in private and in public, childhood cancer was recast as a family of diseases that should be managed by specialists: paediatric oncologists. At the same time, childhood cancer was presented to research sponsors and to the public as potentially curable through the use of chemotherapy. This process can be seen in the USA and Europe, as well as Great Britain, but the circumstances of the British development of the profession are peculiar and worthy of particular focus; as we have discussed, specialised care for children with cancer grew out of paediatrics and radiotherapy in Great Britain, not from research in pharmaceuticals and haematology, as it did in the USA.[2]

The establishment of a nationwide network of specialists who could and would treat the majority of British children with cancer took careful negotiation over a number of years. Large-scale trials of chemotherapy for leukaemic children, designed to recruit hundreds of patients each year, were not launched until 1969, when there was at least some expectation the majority would experience significant benefit in exchange for treatment risks and side-effects. Children with solid tumours, meanwhile, unless they attended the pioneering centres in London and Manchester, were unlikely to receive chemotherapy during the 1960s. Clinicians who wished to offer these children the best available treatments often chose to follow protocols developed in the USA, and some British children were enrolled in trials administered by the big American cancer research centres (see, e.g., Burchenal, 1968). But few children with the rarer forms of solid tumours had access to chemotherapeutic treatment until the late 1970s. Throughout the 1960s and 1970s, the treatment a child with any form of cancer was offered depended heavily on where she or he lived: those near a specialised cancer centre that maintained good relationships with referring physicians would be given experimental drug regimes in hopes of a few extra months of health; those not so well placed might have blood products, steroids, perhaps surgery and radiotherapy, and would spend less time living in the hospital.

During the late 1960s and 1970s, however, care for British children with cancer became increasingly standardised as more local hospitals referred cases to an expanding network of specialist centres. The clinical trials programme was extended to encompass every child, first those with leukaemia and then all children affected by cancer. The proportion of children with leukaemia treated in clinical trials rose dramatically during this period, from around six per cent to near 80 per cent. The number of children with solid tumours receiving care planned by specialists in cancer medicine – as opposed to treatment directed by surgeons – also rose considerably over the course of the 1960s and still more in the 1970s as, through their publications, the new experts in childhood tumours grew in number, confidence, and stature, and began to demonstrate significant improvements in outcome for some tumour types. Full-time posts in paediatric oncology were created in the larger, research-focused children's hospitals, and the term 'paediatric oncology' came to mean a career path, as well as a field of specialised medical practice.

Professionalisation and the centralisation of care were interlinked in complex ways. For doctors to define themselves as paediatric oncologists, they needed to be seeing the majority of children diagnosed with cancer, and to be viewed as the best specialists to treat children and research paediatric cancers, even while other research communities, such as epidemiologists, maintained an interest in this patient group. When dealing with a set of conditions as rare as childhood cancer, professionalisation required centralisation; if the pool of available patients was split between different medical parties then clinical trials and the accumulation of expertise would proceed too slowly and no special expertise could be demonstrated.

Bringing Leukaemia and the Solid Tumours Together

In the USA, from the 1950s, hospitals specialising in chemotherapy research treated children with solid tumours on the same wards as those with leukaemia. In Boston, Farber gave promising new drugs to children with tumours that had metastasised, reasoning that all dispersed malignancies were fundamentally alike in their resistance to the localised treatments of surgery and radiotherapy, and their sensitivity to chemical attack. In the laboratories of the Burroughs Wellcome Company, George Hitching and Gertrude Elion were synthesising antimetabolites tested in rodents at SKI; their top performers were given to all children with inoperable cancers in Memorial Hospital from 1953, together with courses of steroids, also only recently introduced to medical practice. Their key clinical partner in this programme of experimentation was Joseph Burchenal. From its foundation, the paediatric ward at the M. D. Anderson Hospital in Houston saw all manner of childhood malignancies, and worked with hospitals across the region to support the care of children with a wide variety of cancers.[3] As we saw in Chapter 3, this spirit of experimentalism – employing children as test subjects for chemotherapeutic agents with little hope of cure – was anathema to many British paediatricians.

Only the children's hospital in Manchester treated all varieties of tumour alongside leukaemia during the 1950s. Everywhere else, leukaemia sufferers were treated by specialists in blood diseases and biochemistry, and those with solid tumours by surgeons and radiotherapists. Records for how children with cancer were treated in the 1950s and 1960s are scattered, and preserved material often closed to study under legal rules to protect the privacy of patients. However, the process by which the two sets of interested clinicians came together in the 1960s can be mapped. This meeting of interests was often brokered by generalists within the new specialism of paediatrics or by pathologists with a broad interest in paediatric cellular abnormalities, who believed and publically argued that for biological reasons it was beneficial to focus clinical and pathological studies of malignant processes around the status of childhood cancers as paediatric, rather than their physical locations. Solid tumours grew and metastasised much more rapidly in children than they did in adults, and as paediatric radiotherapist Ivor Williams had shown in his survey of childhood malignancies in the 1940s, many of the cancers comparatively common in the young were extremely rare or never seen in adults (see Chapter 1). During the 1960s, clinicians and researchers attempting to persuade funding bodies and hospital departments that studies of the diverse childhood cancers belonged together successfully argued that unravelling the natural history of one disease might elicit the origins and progress of another – as appeared to be the case with Burkitt's lymphoma and acute lymphoblastic leukaemia – and that treatment effective against cells in one cancer might be efficacious against others.

In Great Britain as in the USA, the threat of rapid metastases – often before any lump from a primary tumour could be detected by eye or hand – forged a strong link between the blood cancers and paediatric solid tumours. As almost all diagnosed cases of paediatric solid tumours were assumed to be metastatic by the time they reached medical attention, treatment had to be more than local. Chemotherapy was therefore a natural choice.

Clinicians and researchers treating children's cancer as a unified field of study made one further, pragmatic, argument for their strategy: these conditions were extremely rare. Even in a large hospital, expertise in a single childhood cancer was insufficient to justify a specialist doctor, let alone a specialist unit properly equipped to support patients undergoing intensive chemotherapy. Some even argued that expertise in an individual childhood cancer – save for perhaps acute leukaemia – could never be acquired in a single geographical region: one would not see sufficient cases in the course of a career to develop an in-depth understanding of the disease process and the extent of variation between children with 'the same' cancer. Thus, to be a specialist in *any* childhood cancer, it was necessary for paediatricians to specialise in *all* children's cancer or work in a team that together could treat all forms. This pragmatic argument played a crucial part in the creation of the first hospital posts for 'paediatric oncologists' charged with treating all children with cancer in their region and teaching new generations of doctors how to do likewise.

Professionalisation Around the New Problem

American specialists in childhood cancer, keen to explore chemotherapy's potential as an alternative or supplementary treatment modality
across a number of tumour types and blood cancers, lobbied for support from the NCI, colleagues, hospital administrators, and families
of sick children.[4] In Great Britain, the grouping together of children
with diverse cancers was more gradual, built through the mediation of
determined pathologists, radiotherapists, and in some cases nurses, who
directed parents of children discharged by surgeons towards proponents
of chemotherapy treating leukaemics.[5] The resulting history is one of
contingency and personality. As we saw in Chapter 1, determined and
curious individuals who find like-minded colleagues in other disciplines
can act as the nuclei around which specialised and integrated services
grow. Edith Paterson in Manchester and Ivor Williams in London initiated habits of working between disciplines and hospitals that continued
to develop through the 1950s and 1960s. Children with solid tumours
were admitted to clinics administering chemotherapy earliest in these
two cities, where a model of centralised care for children was already
well established.

London

St Bartholomew's Hospital, as outlined in Chapter 1, was well known
for integrating laboratory studies and ward practice, and therefore
an obvious choice when the Medical Research Council (MRC) was
seeking centres in which clinical investigations of scientifically interesting diseases could be conducted. The radiotherapy department of
St Bartholomew's, within which many of the leading figures in the field
had trained, was especially highly regarded. Ivor Williams had been
conducting research into children's cancers since the 1930s; he led the
care for child patients through the 1940s and 1950s, and in the 1960s
continued to draw visits from practitioners in other centres seeking
experience with young patients. Williams' reputation as both a paediatrically minded therapist and a leading researcher in cancer therapy also
established his hospital as a popular choice for doctors and surgeons to
send their young cancer patients for treatment. From 1960, for example, all Great Ormond Street's patients with abdominal tumours were
referred to Williams' care for the radiotherapy part of their treatment
(Bond, 1975, p. 1202). Williams also had charge of a ward for children
and young adults suffering from Hodgkin's lymphoma, a disease against

which he was having considerable success through the use of wide fields of high-dose radiotherapy.

In the 1960s, however, care of children – and adults – with cancer attracted the research attention of the hospital's highly successful haematology department. As we have discussed, some of the most influential chemotherapy researchers in the USA had extended their interest from blood cancers to solid tumours, and this same pattern occurred in London, where haematology services expanded to cover the application of chemotherapy to a wide range of patients suffering from cancer. One of St Bartholomew's most senior clinicians was Sir Ronald Bodley Scott, an internationally renowned haematologist and specialist in acute leukaemia. His research into blood diseases received ongoing financial support from the MRC, and his stature and successes attracted many ambitious new haematologists to the department. Haematology was an exciting young speciality in the 1960s, when recent research into anaemia, haemophilia, and other clotting disorders had led to detailed understanding of many of the most common or serious blood diseases, and to a range of effective treatments.

In the mid-1960s one of Bodley Scott's registrars – those training to take up posts as senior academics or hospital consultants – was Jim Malpas.[6] Malpas had trained at St Bartholomew's, then worked as a lecturer in the Nuffield Department of Medicine at Oxford while conducting research into vitamin B12 and iron metabolism. Becoming interested in Scott's first trials of methotrexate, Malpas imported the drug from the USA to administer to adult and child leukaemic patients. The paediatricians in charge of children with leukaemia at St Bartholomew's, however, were reluctant to allow this on the grounds that children invariably relapsed. Chemotherapy appeared to put families through further agony: remissions put parents through the grief of losing a child to the illness twice over; at diagnosis and then at relapse.

Scott and his team had strong links with chemotherapy researchers in the USA, meaning they could access experimental agents and protocols detailing how they should be administered. Scott's previous senior registrar, Gordon Hamilton Fairley, maintained correspondence with centres in the USA reporting successes against acute leukaemia and Hodgkin's lymphoma through the use of intensive cocktails of chemotherapeutic agents, and investigating chemotherapy regimes for use against tumours in adult patients as well as children (Löwy, 1996, pp. 60–1). In the mid-1960s, American centres reported marked improvements in cure rates over what could be expected through surgery and radiotherapy alone for several childhood tumours, including

Wilms' tumours and cases of advanced Hodgkin's lymphoma (D'Angio et al., 1980).[7] And work on non-Hodgkin's lymphoma (NHL) in children suggested that treating it akin to acute leukaemia would also result in dramatically improved outcomes.[8] At St Bartholomew's these research findings were translated into treatment plans for adult cancer patients as far as possible, but chemotherapy was not administered to children with solid tumours.

Despite difficulties in securing a supply of all the necessary drugs and assuring the required level of supportive care, by the end of the 1960s Fairley had developed a modified form of the American protocol for Hodgkin's lymphoma that could be administered to British adult patients. In the very early 1970s, Fairley introduced Malpas to Ivor Williams, and suggested Malpas take over the management of children arriving at St Bartholomew's for radiotherapy and incorporate chemotherapy into their treatment plans. These children fell into three categories: new admissions; those referred from Great Ormond Street; and those referred from other parts of the country where doctors had heard St Bartholomew's specialised in children's cancer and thus could offer integrated care. Malpas could not include in his small-scale trials those from Great Ormond Street, as they remained under the care of its specialist surgeons, whose assistants visited to ensure coordinated treatment and a smooth passage of information on patient progress. The other two groups were suitable for inclusion and Malpas began to apply the American chemotherapeutic regimes to these challenging cases.

Realising that treatment centres needed to cooperate to attain large enough patient groups, Malpas met with Tim McElwain, an oncologist at the Royal Marsden who had trained in cancer medicine at St Bartholomew's and was also treating a few children according to American protocols. They decided to form a Children's Solid Tumour Group (CSTG) in the early 1970s.[9] The two hospitals conducted studies on the rare Ewing's sarcoma of the bone, the lymphomas, and rhabdomyosarcoma, a tumour originating in soft tissue, usually muscle. The last of these began to appear curable once chemotherapy was added to the standard treatment plan of surgery and radiotherapy.

By 1973, Great Ormond Street had established two consultant posts for the management of children with cancer. Judith Chessels, a haematologist taking over some of the laboratory work of Roger Hardisty, was organising the treatment and care of children with leukaemia, with the additional bonus, or burden, of responsibility for allocating beds in the hospital. She and Hardisty persuaded the Leukaemia Research Fund (LRF) – closely allied to Great Ormond Street as discussed in

Chapter 5 – to supply initial funding for the appointment of an additional consultant responsible for extending the developing haematology service to the care of children with solid tumours. This was to replace the system where children with solid tumours were placed under the care of surgeons in the hospital and sent to St Bartholomew's for radiotherapy, and to supplement the existing staff devoted to ensuring effective communication and cooperation between the sites. Since the early 1960s, Jane Bond had been working as a 'tumour research fellow', monitoring patients' follow-up therapy and progress over the years, and supporting those surgeons who saw children with tumours not located in brain or bone. Children with bone tumours were typically referred to the Middlesex Hospital, which specialised in cancers of the skeleton, while children with brain tumours were usually seen by neurosurgeons within Great Ormond Street before being sent to University College Hospital for radiotherapy. The creation of the additional post at consultant level made it possible to establish a new outpatient service for chemotherapy and to recruit these surgical patients into it (interviews with Judith Chessels, 2004, and Jon Pritchard, 2004).[10] Pritchard and Bond quickly joined the CSTG, adding their patients to the collective pool of trial subjects and sharing the task of keeping abreast with research.[11] Judith Chessels has reflected on the significant difference this reorganisation made. Previously, only children with leukaemia had been offered any anticancer drugs:

> The chemotherapy service was really just for the haematology service. And it was only really when Jon Pritchard came as consultant oncologist, and we started to pull patients away from surgeons, that it got more organised. That took quite a while. It was really a political thing (interview with Judith Chessels, 2004).

In London, with so many powerful teaching hospitals and influential clinicians and researchers, the traditional divide between the blood and solid cancers only gave way to a model of unified cancer services gradually.[12]

The London group realised that children were being treated for cancer all over Great Britain, but the quality and intensity of this treatment varied considerably. While some were treated in centres following the latest ideas about what constituted best practice, others, often in adult cancer wards with no psychological or social support for patients and parents, were still receiving treatment designed for fully developed bodies, not growing children. The London paediatric oncologists were convinced a nationwide study group was necessary if the situation was to be rectified. Paediatric oncologists in the North West of England were also forming a local professional network, and considering how to extend their knowledge of rare cancers, trial designs, patient care and new chemotherapy agents, to cover all affected children in the region. These groups' ambitions reflected those seen in the American leading centres, but also depended heavily on the nature of the NHS: government funded treatment, it could be reasoned, should be the same for patients everywhere.

Manchester, Birmingham, and Liverpool

The special circumstances around care for children with cancer in Manchester have been described in earlier chapters. The Royal Children's Hospital at Pendlebury continued to take a pioneering role in developing new approaches to standards of treatment for childhood cancers in the 1960s, through its ongoing collaborations with the Christie Hospital and the medical school of the Victoria University. The tumour registry established by pathologist Henry Marsden and paediatrician Edith Paterson had, indeed, led to increased referral rates of childhood cancer cases from across the whole of North West England.[13] The registry further strengthened the unit's reputation as a centre of excellence, as clinicians in Pendlebury used it to collate evidence about childhood tumours' incidence and responsiveness to different forms of treatment.

As discussed in Chapter 2, Henry Marsden had initially set up the tumour register after spending a year studying alongside Sydney Farber in Boston, shortly after the publication of Farber's startling results from administering chemotherapy to children with acute leukaemia. The task of running the register, coordinating pathological analyses of new samples, and servicing requests for information from the data collected was quickly recognised to warrant someone's full-time attention. This responsibility fell to Jake Steward, a paediatrician in the children's hospital whose own health limited his ability to work directly with patients. In the mid-1950s, Steward began compiling detailed analyses of incidence patterns and eventual patient outcomes, with some expectation this might lead to a worthwhile publication after the first decade of the register's existence.

Since 1950, the professor of paediatrics in Manchester, Wilfrid Gaisford, had been committed to research into childhood cancers and to improved and integrated services for patients. In 1963 he was appointed as Chair of the paediatric subcommittee of the MRC's working party on leukaemia trials. This post carried with it sufficient MRC funding to expand the Manchester team engaged in clinical experimentation in children's cancers, and facilitated the hiring of an additional paediatrician with special responsibility for those patients with leukaemia or solid tumours being treated at Pendlebury (interviews with Pat Morris Jones, 2004, 2005).

The physician appointed was Pat Morris Jones, who had earlier worked as a Resident Clinical Pathologist in the pathology department under Marsden and as a tutor in the Department of Child Health with general paediatric duties. In her new post, Morris Jones set about developing centralised care and specialist services for the full range of paediatric malignancies in the region, a process she had completed by the end of the decade. She worked closely with Dorothy Pearson, the radiotherapist at the Christie who saw most children requiring radiation treatment and ensured that patients sent to the Christie were also seen by staff at Pendlebury. The respected paediatric surgeons employed at Pendlebury helped establish an excellent surgical reputation across the region, leading to more solid tumour cases being immediately referred there, rather than operated on by adult surgeons in general district hospitals. Integrated treatment plans, at times involving the administering of chemotherapy before an operation, were being implemented by the Manchester clinical team by 1970 and made available to the majority of children in the North West region by the mid-1970s (ibid).

During the 1960s and early 1970s, children with solid tumours living in or near Liverpool were looked after in the large Alder Hey Children's Hospital in West Derby and the old Royal Liverpool Children's Hospital on the other side of the Mersey.[14] Paediatric oncologist at Alder Hey, John Martin, delivered treatment to children in both centres and began using American protocols in 1971. Centralisation in cancer services for children was also taking effect in the Midlands. In the late 1960s, Jill Mann was expanding the haematology service in Birmingham's Children's Hospital to encompass oncology. This was one of the first specialist children's hospitals to be established in Great Britain; it had also been the first hospital outside of London to boast a Chair in Child Health.[15] By 1970 the hospital was running small trials of experimental combination therapies, including chemotherapy, for some children with solid tumours.

Clinicians from Manchester, Birmingham and Liverpool met frequently to discuss difficult cases and share their literature surveys, as did the cluster of paediatric oncologists in London. The northern group also met clinicians and researchers at oncology meetings held in other countries; Morris Jones, Pearson, and Mann were among the earliest members of the *Société Internationale d'Oncologie Pédiatrique* (SIOP), an international society for paediatric oncologists initiated by Odile Schweisguth in 1969.

SIOP had itself grown out of an informal club for paediatricians, surgeons, pathologists, and others interested in childhood cancers, who met for the first time in 1967 in the paediatric department of the Institut Gustave-Roussy in Villejuif near Paris, France, where Schweisguth had been Head of Paediatrics since 1951. The group convened in the same place a year later and decided to formalise their gatherings. Thus, in November 1969, at a meeting in Madrid, the club adopted a constitution and became SIOP, with 28 founding members from across mainland Europe. The organisation has continued to expand its region, remit, and interests, under the name of the International Society of Paediatric Oncology, acting as a coordinating body for clinical trials conducted throughout Europe.

Before the inception of SIOP, Morris Jones, Pearson, and Mann, like many of their British colleagues in the fledgling discipline of paediatric oncology, had joined the paediatrics section of the American Society of Clinical Oncology. They attended meetings in the USA, forging contacts with doctors in the leading research centres there and internationally, and gaining access to the latest trial protocols and experimental drugs. At this time, however, there was no national organisation and no national voice for paediatric oncology in Great Britain. Neither were there routes into the specialism for young physicians trying to acquire expertise. Training fellowships in 'paediatric oncology' in the large US centres were advertised in the medical press from the early 1970s, and many British paediatricians undertaking the treatment of childhood cancer patients went to these centres for periods of time ranging from three months to two years, returning with experience of applying and helping develop new chemotherapy protocols. These fellowships were, in some cases, funded by the LRF or the MRC as a way of bolstering research back home. The reminiscences of physicians who held these fellowships in the early and mid-1970s refer to their sense of having qualified as paediatric oncologists by virtue of obtaining a BTA – a 'Been To America' (Ablett, 2002a; interview with Jim Malpas, 2004; interview with Tim Eden, 2004).

Outside the major British cities, the type of treatment available to children with solid tumours was determined by where they lived, which specialists were known to their general practitioners, or which hospital they first entered. Whether they wished to or not, many families of children with cancer were not able to access the thin network of centres offering the most aggressive treatments and the best chances of cure. By combining all the records of children treated with chemotherapy protocols in hospitals around Great Britain, we calculate that between 1960 and 1965, approximately three per cent of sufferers were treated on clinical trials and a further three per cent received at least some drug therapy from individual physicians operating individually (FD 7/323; FD 7/325; FD 7/330). Even in the case of acute leukaemia – for which the MRC had put in place the most substantial trial infrastructure – only 17 per cent of children diagnosed with the condition between 1963 and 1967 were treated by physicians partaking of the MRC programme, and not all of those received the suggested drugs and/or recommended doses (Medical Research Council, 1971). Results published by clinicians running trials that used only parts of the MRC drug protocols fell short of those achieved elsewhere (Thompson and Walker, 1962).

Calls for the Centralisation of Services and Appeals to the Hope of a Cure

As those treating childhood cancer came to call themselves 'paediatric oncologists' and formed affiliations for sharing expertise and lobbying for greater resources, supplies of one resource were of particular concern: patients. In the late 1960s and the 1970s, paediatric oncologists developed and deployed arguments calling for the centralisation of care across the whole of Great Britain, and for clinical trials of new chemotherapeutic treatments for cancer to be available to children wherever they lived. During this period, British paediatric oncologists began to establish their position within an increasingly international field.

Arguments over the pros and cons of centralisation centred largely on childhood leukaemia; the MRC childhood leukaemia working party, for example, sought to quantify the value of expertise in a 1971 publication by demonstrating that children treated solely by specialists lived, on average, more than twice as long as those seen by nonspecialists (Medical Research Council, 1971). But the same points were also made regarding those tumours not in bone or brain, often with the explicit reasoning that any observable benefits for leukaemic children could be expected to extend to children with other forms of cancer. Some contributors to the debate took the unity of paediatric malignancies to be self-evident. When the connection was argued explicitly, advocates for centralisation asserted that paediatric malignancies required systemic treatment with chemotherapy to prevent the growth of undetectable 'micro-metastases', or emphasised the special emotional and psychological needs of children that required they be cared for together and away from other cancer patients.[16] Perhaps the most persuasive claim used, however, was that treating childhood cancer cases alongside and in the same way as cases of leukaemia would yield similar dramatic improvements in cure rates. The cost of expanding chemotherapy trials, in money and in suffering, was disconcerting to many doctors and to some within the MRC. The possibility of cure, however, made it increasingly difficult for clinicians to refuse to support efforts to deliver curative treatment to all children afflicted by cancer.

Through the 1960s, the NCI's acute leukaemia task force had tested series and then combinations of drugs against acute leukaemia, extending survival for child patients from an average of four to 12 months. But by the middle of the decade, as more children were surviving many months without relapse in their bone marrow, clinicians noted an alarming proportion suffering leukaemic infiltration of the central nervous system (CNS). Such relapses could be treated with chemotherapy injected into the spinal fluid, or radiotherapy to the brain and spinal column, but many went on to suffer relapse within the marrow as well. It appeared the CNS served as a 'sanctuary site', affording leukaemic cells a place to hide from the effects of drugs administered through the blood, drugs that could not effectively cross over into the brain to kill the cancer cells within.

The MRC trials of chemotherapy for leukaemia running in the same period, the mid-to-late 1960s, had yielded only small improvements in survival figures and it was still the case for the majority of patients that British experimental treatment merely prolonged life, it did not avert death; only three and a half per cent of those treated by the MRC were surviving for five years (see James and Kay, 1967; *The Lancet*, 1971; *The Times*, 1973).[17] Many researchers preferred to measure the effectiveness of treatments by studying '10% survival times' – the length of time by which only ten per cent of trial participants were still alive – rather than the proportions surviving for five years or longer. The numbers reaching this latter milestone were infinitesimal and there was speculation the disease process might actually be following a different biological line of development in those reaching this point.[18]

In 1967, a group of clinicians at St Jude's Hospital in Memphis, headed by Donald Pinkel, had begun to administer radiation to the head and spine of child patients with leukaemia under the assumption there would be undetectable leukaemic cells lurking within the CNS from the outset. Pinkel had been appointed as research director for the hospital's opening in 1962, on the strength of his work at Roswell Park in Buffalo.[19] His programme of trials was presented in the medical literature as a conscious attempt to not only extend survival, but effect a 'permanent' (Hustu et al., 1973)[20] or 'deliberate' (Spiers, 1972)[21] cure. Promising results swiftly started to emerge, with fewer children relapsing in the CNS. In 1972, Pinkel and his colleagues announced that half of those children treated five years earlier had survived without relapse. They reasoned that for many patients this relapse-free period would turn out to be a reliable indication of cure (Pinkel, 1972).

Intense chemotherapy carried a high risk of distressing side-effects, such as inflammation and infection of mucous membranes, affecting the ability to eat or digest; damage to the brain, to bones, and to hearing; and even death from infection or haemorrhage. As we have seen, British clinicians treating children with cancer were typically trained as paediatricians first, cancer specialists second, and many struggled with the idea of subjecting children to such grievous treatment courses for the small chance of a few extra months of discomfort for the child and fear for the parents. But the possibility the suffering might yield a cure changed their risk–benefit analysis. Aggressive treatments seemed warranted to a wider cross section of the medical profession if 50 per cent of children had a chance of cure. The Memphis team's announcement stressed that childhood acute lymphoblastic leukaemia 'can no longer be considered an incurable disease. Every child with this disease deserves the opportunity for permanent cure' (Rhomes et al., 1972, p. 390). This right of the leukaemic child to be given the chance of cure became the dominant trope in British calls for professional organisation and better state funding for paediatric cancer services over the next five years, and was still being employed in public appeals for charitable giving during the 1980s.[22]

Given the early reports that prophylactic CNS treatment reduced the risk of relapse, the MRC childhood leukaemia working party followed 'the St Jude template' when constructing the 'UK ALL' trial series, launched in 1970 (Lilleyman, 2003, p. 46). These trials relied upon clinicians competently distinguishing the more treatable lymphoblastic form of acute leukaemia from the myelogenous form and even rarer varieties affecting other cell types; the typing committee discussed in Chapter 3 had successfully disseminated techniques for minimising the number of unclassifiable cases, and cell visualisation technologies had developed considerably. The childhood leukaemia working party abandoned its earlier trials programme, in which drugs had been administered in series, to run large-scale trials of intensive regimes on the American model. In order to recruit sufficient patients to yield rapid results, the National Health Service (NHS) had to commit to funding a comprehensive array of regional specialised centres, equipped to safely administer the chemotherapy as specified.

Acute leukaemia was the first paediatric malignancy to attract the resources necessary to develop a curative regimen and make this available to all patients across Great Britain, but the solid tumours were soon to be included in this vision. Calls for centralisation, however, provoked disagreement between the diverse groups of clinicians, and a few epidemiologists, who claimed childhood cancer as their area of enquiry. To achieve funding for centralised clinical services, for leukaemia and ultimately for all childhood cancers, all interested experts had to be brought on board the campaign to recast childhood cancer as curable and appropriate for treatment by oncologists.

In July 1968, *The Lancet* published an article entitled 'Cancer in Children', a response to the publication of the Manchester textbook, *Tumours in Children*, edited by Henry Marsden and Jake Steward (1968a). *Tumours in Children* had demonstrated that cancer had moved from third to second most common cause of death in British children over the age of one year and laid out the case for planning special services for children with tumours. *The Lancet* article backed this call to arms, stating bluntly that the best hope for any child with cancer was to be immediately referred to the care of specialists with extensive knowledge of childhood cancers. Data collected from the Manchester Children's Registry, and from Alice Stewart's group researching childhood cancer incidence, were brought together to strengthen the argument that centralisation of care could only improve the chance of good outcomes: while cases were being seen by clinicians witnessing, on average, fewer than one case per year, there was no hope that children would receive the best possible treatment.[23]

The Lancet article spoke optimistically of the potential of chemotherapy to render more cases of childhood cancer curable, but, as the author immediately pointed out, patient supply was imperative:

> new agents can be properly evaluated only where sufficient patients can be seen and treated by staff with a thorough knowledge of childhood cancers and their natural history.

The article concluded that, given the rarity of many conditions, 'the answer must lie in centralised resources'. Such a system of large centres would deliver better epidemiological material as cases could be identified by those more skilled at histological diagnosis. The piece closed with a bald statement that appeared to imply studies such as Stewart's were less important than registries such as Marsden and Steward's:

> The time for epidemiological inquiry is when the child's cancer is diagnosed and not when death certificates are filled.

The Lancet article led to a flurry of decreasingly polite correspondence about centralisation, and coverage by *The Guardian* newspaper, in which assertions and counter-assertions were made on who was most suitable to manage children with cancer. Alice Stewart and Edith Ledlie from Oxford's Social Medicine research group argued against what they saw as a call on the Department of Health to fund new specialised units, and criticised the author of the original article for ignoring the role of the Oxford Survey as 'a source of published information in connection with the problems being discussed': namely, how better to collect and classify cases in order to uncover the causes of childhood cancer. They drew on the Survey's records to demonstrate that the type of hospital in which treatment was administered had no impact on prognosis for children with Wilms' tumour, and used this as evidence for a counter-call: that any available additional funds should go to the Oxford Survey, then substantially dependent on contributions from the American Public Health Services (Stewart and Ledlie, 1968).[24]

In reply, Marsden and Steward countered that the building of new wards was not required, existing bed spaces should merely be brought together within regional centres. The real point of contention, however, was the relative importance of clinical practice and the study of patient data. Usual British reserve was abandoned in the closing paragraph of Marsden and Steward's (1968b) letter:

> It is difficult to see what conclusions can be drawn from the Wilms' tumour survival-rates cited by Dr Stewart and Dr Ledlie. Are they meant to prove that clinical experience in paediatric oncology does not, and never will, matter? ... Common sense tells us that greater centralisation makes for more experienced clinicians and the better use of available therapy. Epidemiological investigations have their limitations and are no substitute for a clinical service designed not only to care for the children who have cancer now, but also to provide better treatment for those who will develop the disease in future.

Marsden and Steward renewed their calls for clinical organisation in order to deliver expert care by 'specialists who are particularly experienced in paediatric oncology', to assess properly modifications in treatments and attract researchers into the field.[25] They took as axiomatic that the most interesting research into childhood cancer was to be done on the living not the dead, and in the realm of treatment not aetiology. They also assumed that cooperation across hospitals and between specialist clinicians would facilitate clear results in trials of chemotherapy for solid tumours, as it had for leukaemia. And they argued for centralisation with passionate faith that it would deliver better outcomes, improvements beyond the evidence they had marshalled for extended survival times and cure rates.

Larger-scale trials were advocated as the means of ensuring clinical experimentation in childhood cancer would be properly scientific. The market-leading paediatric oncology textbook of the 1970s called on its readers to help establish paediatric oncology on as firm a scientific footing as that seemingly enjoyed by paediatric haematology, proven by the scale of the NCI's leukaemia project:

> As a result of the years of experience provided within the atmosphere of the collaborative group, it is obvious that the era of empirical therapy is dissipating ... The pediatric oncologist is now deeply committed in a dynamic period of experimental and investigational therapy that promises even greater benefits within the foreseeable future (Sutow et al., 1973, p. ix).

American researchers had published articles and held conferences on how to design trials to deliver clear results on the comparative efficacy of different treatment options. British clinicians aspired to meet these standards.

Centralisation in Practice in the 1970s

British paediatric oncologists argued that more children would be cured if a pool of specialist centres across Great Britain was established as part of an international network of clinical research. The promise of cure would justify the additional expenditure required for intensive chemotherapeutic treatment and persuade other medical practitioners to refer on cases of malignancy. All child patients would be enrolled in clinical trials administered by cooperative study groups.

In the mid-1960s, a nationwide cooperative group had been established in the USA to run a trial of the various chemotherapy regimes in use across the country for Wilms' Tumour, comparing one with another and any chemotherapy with none. The results were clear: chemotherapy made the cancer survivable for a much larger proportion of patients, even those with disseminated disease. Applying this American research, an MRC working party on embryonal tumours opened a British trial for Wilms' tumour patients in 1970. When the results were published in 1978, replicating the American findings, the group responsible for the trial proudly stated they had treated 57 per cent of British children suffering from that form of cancer who were eligible over the period of the trial, and 31 per cent of all those diagnosed with the condition.[26]

In light of the results achieved for patients with Wilms' tumours and for children with other tumours treated in small-scale trials according to American protocols, the United Kingdom Children's Cancer Study Group (UKCCSG) was founded in 1977 to extend chemotherapy research to the paediatric solid tumours. This small group, initially with only 14 members, was the result of the London CSTG meeting up with clinicians from the northern triangle, and from hospitals in Bristol and Newcastle, at the Children's Hospital, Birmingham.[27]

Over the first few years the study group encountered substantial resistance, only managing to register half of new cases of childhood cancer.[28] Members of the group campaigned vigorously and successfully to improve this referral rate. They claimed to demonstrate that specialist care delivered better patient outcomes; at times, when children in relapse were referred following local nonspecialist treatment, it would become clear that blood levels had not been monitored, and any continuation therapy administered had been inadequate.[29]

The MRC working parties and the UKCCSG collaborated with the Childhood Cancer Research Group based in Oxford – the organisation that replaced Alice Stewart's Oxford Childhood Cancer Survey in 1975 – in the monitoring of childhood cancer epidemiology. Together, the three research groups coordinated the development of clinical trials for British children with cancer, centralised registers of cases, and standardised diagnosis and treatments across Great Britain. The three bodies differed in their stated aims, but managed to negotiate working relationships that ensured the work of each was seen as complementary, justifying the expensive provision of paediatric cancer services required by their research programmes.

Centralisation and Cure

As the profession of paediatric oncology took shape and a tertiary referral scheme was implemented in Great Britain for children with cancer, published accounts of research and appeals for support used the notion of cure – increasing the chance of, and extending that chance to all children – to demonstrate the value of the new field of clinical investigation and the reorganisation of services. The notion of 'cure' was a difficult one in the case of children treated for cancer. For adult cancer patients, 'cure' has often been spoken of once a patient has been clear of cancer for five years. No single approximation has held within paediatric oncology, partly because the time taken for the risk of relapse to drop to a few per cent varies widely between cancer types, and partly because death from cancer's effects after five years would still constitute an enormous shortening of life expectancy. In the papers of the 1970s, authors were careful to present their definitions of 'cure' as working definitions (*The Lancet*, 1968).[30] In Chapter 8 we discuss the problems of defining 'cure' for childhood cancer survivors in more depth, and examine various attempts to find a solution.

Dramatic increases in survival rates and durations were seen in small-scale trials, in the USA and in Great Britain, as a result of the introduction of intensive chemotherapy protocols. The steady flow of reports of progress in the battle against lost years of life brought the (in-) 'curability' of childhood cancer into question. By the end of the 1970s, childhood cancer had been reconceptualised as a set of conditions that were increasingly curable through – and only through – rolling programmes of clinical trials, necessitating the gathering together of all those suffering from the illness under the care of specialists in paediatrics, as well as cancer. The majority of children in Great Britain, and in North America and Europe, with leukaemia and solid tumours not in brain or bone were being seen by specialists in children's cancer who prescribed long courses of chemotherapy, often in conjunction with surgery and radiotherapy. This approach was presented and accepted as the new norm for those within the medical profession and for the public at large. Stories of children battling cancer had become common in the media and in literature, broadcasting the message of curability, in marked contrast to the situation 20 years earlier.[31] While many children still perished, the fact that a substantial proportion could survive into adulthood made childhood cancer more palatable to the public, and made hopeless cases appear all the more tragic.

The UKCCSG and the Benefits of Specialist Care

Substantial improvements in patient outcomes were achieved by the British paediatric oncology profession for all forms of childhood cancer, as the UKCCSG united geographically dispersed specialists, secured dedicated funding for research, and brought experimental treatments from around the world to British hospitals. Minutes of the group's meetings reveal the multifront approach of the campaign to increase resources for the specialism of paediatric oncology, and the dedication of members to advancing on each front as speedily as possible. While many of the group's first members treated children suffering from leukaemia, as well as those with tumours, for its first 15 years, the group's experimental work was all focused on solid tumours, as trials for acute leukaemia were already well established through the auspices of the MRC. Paediatric oncology in Great Britain remained firmly rooted in research into paediatric tumours; professional relationships with those in the field of paediatric haematology, and with the MRC working party, had to be established and nurtured before the UKCCSG could also input to discussions about leukaemia clinical trials programmes.

During 1977, the group met every two months, rapidly expanding by inviting representatives from hospitals where it was known children with cancer were being seen by clinicians with an interest in chemotherapy.[32] The group immediately established links with the Oxford Survey, and a registry of children's tumours was established – initially based at Alder Hey hospital – that both organisations could access for research into cancer pathology and incidence.[33] From the inaugural meeting of the group, members also began to debate and design treatment protocols and methods of testing these for NHL and for neuroblastoma, a tumour that develops from cells in the endocrine and nervous systems; the NHL trial was enrolling patients by the close of 1977.[34] By the end of its first year, 296 children had been registered by physicians in membership: 115 suffering from leukaemia, 34 with tumours of the CNS, 25 with neuroblastoma, 24 with NHL, 19 with Wilms' tumours, 17 with rhabdomyosarcoma, 19 with Hodgkin's lymphoma, and 63 with other tumours.[35]

Numbers of patients were small for most disease types. As clinical trials for many diseases required that treatment be differentiated according to the stage a cancer had reached, conducting tests of chemotherapy that could yield significant results in a small number of years was felt to be impossible for the less common conditions unless patients were pooled across countries. The UKCCSG commissioned small groups of members to investigate existing trials recruiting patients or under development in the USA and Europe, and make recommendations as to whether British centres should seek to gain access to those protocols or devise their own trials. Considerable attention was given to analyses of chemo-sensitive brain tumours in 1978, with reports of trials organised by coordinating groups in America and by SIOP. Members could not agree whether to enrol patients onto either, or run a number of small tests of various protocols already being investigated using historical series of patients as the control set against which to measure the efficacy of therapies. There was additional concern that any radiotherapy administered alongside chemotherapy in trials for brain tumours and Ewing's sarcoma, would probably be impossible to standardise across hospitals, as radiotherapy departments – often controlled by professionals outside the sphere of influence of UKCCSG members – had varying equipment and differing notions of what constituted exemplary treatment for these conditions.[36]

Patient numbers for trials were of equal concern to the American cooperative groups and SIOP. UKCCSG minutes document a steady stream of requests from both regions for British patient enrolment in order that trials could deliver answers about best treatment options. By late 1978, the rhabdomyosarcoma trial being run by SIOP was stalling owing to low uptake, and the organisation asked for UKCCSG input.[37] The MRC was also keen to pass the responsibility for trials on solid tumours to the UKCCSG soon after the group's inception. The two organisations began discussing how this might be effected in early 1978, and within two years UKCCSG was the sole body in Great Britain administering clinical trials for solid cancers in children.[38]

Gaining access to sufficient patients continued to be a concern. After three years of operation, over 1,300 children had been registered, and their details entered into a computer.[39] But the UKCCSG had cause to reprimand some of its members for failing to bring in new cases. The minutes of the summer meeting of 1979 record that in order to maximise trial recruitment and avoid potentially skewing patient samples, all centres should be reminded of the need to operate together:

> while most centres were entering the majority of their eligible patients to UKCCSG studies, some had a poor record in this respect and one large centre had not entered any cases at all. It was agreed that the original members of the group had expressed their intention of working together to undertake clinical studies and that, while participation remained voluntary, this was still our prime objective.

Therefore, it was agreed that the Secretary should write to all members to discover their intentions with regard both to the registration of patients and to their entry to the studies.

The hospital in question – the Royal Marsden – was quickly brought into line, faced with the potential loss of access to the protocols being developed by the UKCCSG, to which only members were entitled, access that made it possible to source the required drugs at significantly lower cost thanks to pharmaceutical company sponsorship of the trials.[40] The UKCCSG maintained that clinical studies could only be completed in a timely manner with due randomisation if *all* British children presenting with the relevant cancers were treated according to the one set of trials.

An organisation with the ambition of the UKCCSG required secure funding. In its first years of operation, the group was dependent on support from pharmaceutical companies. Lederle, Eli Lilley, and Wellcome all contributed with direct donations and through meeting the costs of staging symposia or putting out publications.[41] By 1980, however, it had become obvious that greater amounts of reliable funding would be required over a number of years, and an application was successfully made to the Cancer Research Campaign (CRC).[42] The organisation was able to hire a research officer and a clerical assistant, but worries continued about the long-term fundraising strategy the group should pursue, as members doubted the CRC would increase the amounts awarded to support a larger team or extend the funding past the initial three-year period.[43] The CRC, however, communicated to the UKCCSG that it was worth their while asking for five more years of support and for a larger group of staff, if the CRC logo could be included in all UKCCSG letters and publications, and members would produce a leaflet on cancer in childhood that the charity could distribute to the public in a fundraising drive.[44] This arrangement continued throughout the life of the UKCCSG; it is tempting to speculate that the CRC viewed the resourcing of this group as a way to increase public donations, children with cancer having great potential to inspire charitable behaviour.

The UKCCSG wished to be recognised by all relevant parties – within the government and the medical profession – as the seat of expertise in paediatric oncology: the organisation that should be consulted over strategy for funding specialised units in hospitals; over developing training for physicians and for nurses; and for deciding which international protocols to join and which to seek to better through their own designs. In 1983 they faced a challenge to their authority when the Cromwell Hospital, a private hospital in London, issued a statement declaring it had facilities for a paediatric oncology clinic and for performing bone marrow transplants. The minutes record that:

> As no Paediatric Oncologist or Paediatrician known to the Group was involved in this venture it was agreed that we should express our considerable concern with regard to its development...

A letter was duly written to the British Paediatric Association (BPA) to request assistance in preventing the venture succeeding; it is currently the case that in Great Britain, private treatment is not advertised for children with cancer.[45]

By 1986, the UKCCSG could, indeed, claim to be the voice for the profession in Great Britain. The group began work on a paediatric oncology strategy, surveying all existing specialist units to map their staffing, facilities, and funding, and to identify deficiencies. Representatives held meetings with the BPA and with the Royal College of Nursing (RCN) to present a coordinated recommendation for the development of paediatric oncology services.[46] The end result, *Cancer Services for Children*, was distributed to the Department for Health and Social Services, the BPA and the RCN, and the Royal College of Physicians in the summer of 1987.[47] It caused enormous media interest, leading to the production of a current affairs programme on the subject.[48] The report proved effective in some areas: while it did not immediately lead to increases in funding for services, three years later the UKCCSG was consulted regarding what training was necessary to ensure the long-term viability of the profession, without junior doctors having to travel to the USA or Europe to acquire the expertise required for consultant-level posts.[49]

Early statistical analyses of the results of the UKCCSG's trials programmes, conducted by the Oxford epidemiology group, clearly demonstrated a significant trend towards increased survival rates for a number of tumours between 1977 and 1986.[50] There were NHS hospitals, however, which persisted in 'doing their own thing', much to the frustration of UKCCSG members. Between 1985 and 1989, many letters were written from UKCCSG chairs to those in charge of health services in the Oxford region, which had not appointed a consultant in paediatric oncology and was continuing to refer some cases to Great Ormond Street but treat others according to local policies with suspected poorer outcomes. When an invitation for local physicians taking care of children with cancer to attend UKCCSG meetings was declined, more letter writing ensued, as it was feared that such a state of division might 'promote isolationism'. Every hospital region, it was believed and argued, should commit to resource research into childhood cancer, cooperatively with the rest of the country. The Oxford region relented in 1989, and opened a UKCCSG-affiliated centre in the early 1990s.[51]

The political strength of the UKCCSG continued to grow in the 1990s. By 1992, the Chair could report that:

> On the political front, it was increasingly apparent that other organisations would recognise UKCCSG as the voice of children's cancer services in the United Kingdom. Recognition as such from the BPA seems likely in the near future.[52]

CRC support appeared secure for the long term, and registration rates were continuing to improve: 71 per cent of all children aged under 15 were being treated in UKCCSG centres. This overall rate concealed far higher registration levels in younger children, as only 45 per cent of 13 and 14 year olds were being referred to UKCCSG centres, largely owing to the inappropriate resources for socialising and relaxation these offered adolescents. Many patients with CNS tumours were also being seen elsewhere, primarily in neurology units, leaving only 44 per cent attending paediatric oncology clinics. Discussions with the Royal Colleges and the Joint Council for Clinical Oncology continued, and over a period of five years an examined training programme for paediatric oncologists, recognised by the Royal Colleges, was agreed.[53]

In the 1990s, the organisation also expanded its remit to address identified weaknesses in its services. The needs of adolescent patients were reviewed, an agreement being reached that the UKCCSG would allow entry to its protocols to young people up to the age of 19, and would look to improve ward and clinic facilities to better meet the needs of these older patients.[54] Further effort was devoted to improving links with the MRC working party on leukaemia in order to bring the two bodies into closer professional communication.[55] New working groups were established to look at treatment techniques hitherto not studied, including one working on a register of children who had undergone bone marrow transplants in order to capture any long-term side-effects and thus produce a guide for all members on follow-up requirements.[56] By 1994 it was reported that over three-quarters of all children with cancer were being registered by the organisation.[57] At the same meeting, it was announced that an important milestone had been reached: there were 10,000 survivors of childhood cancer in Great Britain.[58] The UKCCSG's investment in the support of survivors, including the preparation of a guide, *Out of the Wood*, on securing employment and life insurance, and having a family, appeared vindicated (Ablett, 2002a, 2002b). The group launched further initiatives to support developing countries that could not access the latest treatment protocols or afford to treat all affected children, and to forge stronger connections with parent support groups and smaller, local charities that could offer practical help to families with children undergoing therapy.[59]

After almost 30 years of operation, the UKCCSG merged with the MRC childhood leukaemia group to form the Children's Cancer and Leukaemia Group, a single body coordinating the efforts of the paediatric oncology profession.[60] The survival rates for most cancerous conditions that affect children had improved greatly over this period owing to paediatric oncologists' determination to establish the discipline and enable all children to access the most promising therapeutic protocols and the best equipped wards for the highest possible chance of long-term remission and cure.

7

Living With Uncertainty:
Three Patients on Trial

Over the previous six chapters, we have seen childhood cancer develop from a rare and usually fatal condition, deemed of marginal interest within the field of child health, to become a major, but perhaps manageable, threat to life and a site for intense and exciting biomedical research. The treatments and prognosis for childhood cancer changed dramatically in the 30 years between 1960 and 1990, as clinical trials conducted in the USA, Great Britain, and Europe yielded improvements in survival times and cure rates, even for acute leukaemia and rare aggressive tumours. Snapshots of the experiences of some of the British children with cancer being treated in this period were captured in newspapers, as we saw in the case of Edward Burke; others have been preserved in the files of the Medical Research Council (MRC) and the Ministry of Health. Further cases can be found in published memoirs by patients, parents, and doctors, and yet more can be glimpsed in medical journal articles presenting clinical trial results, where the effects of treatment on particularly responsive individuals occasionally warranted extended commentary. Through these stories we can see how innovations in scientific understanding and treatment changed what it meant to live with childhood cancer, not just medically or biologically, but also personally and culturally.

We open this chapter by looking at the very different experiences of two patients given chemotherapy as adjuvant therapy – the first treated between 1960 and 1962 and the second a decade later – and reflect on the degree to which intensive multimodal cancer therapy seemed justified by the chances of remission and its likely length at these two points in time. We then examine the early efforts of observers within medicine and psychology to make sense of what patients and their families needed to best cope with the experiences they had to undergo. During the 1970s, experts argued over whether it was better to be honest with families about the chances of death or protect them from knowing the truth, until it became apparent that sufficient numbers of children were recovering from their cancers that preparations *had* to be put in place to equip the fortunate for long-term survival. Increasing numbers of families faced the challenge of living with uncertainty, not knowing whether to accept the worst or to prepare for a happier future – for cure. Psychologists and anthropologists worked with families to understand the implications of living with the open status of remission – by definition assumed to be potentially temporary – in order to help paediatric oncologists, nurses, and support workers grasp the meaning of childhood cancer as survival rates climbed.

For a third case story, we consider the impassioned memoir of a Cumbrian mother whose son was treated for acute leukaemia between 1986 and 1990. Her anger at the randomisation in clinical trials and the administration of treatments ultimately given in vain rises from the page. At the same time, her arguments show the degree to which, by 1990, orthodox medicine was well established as the locus of expertise in childhood cancer. We close the chapter by placing this case story in the context of ongoing fears of the health risks posed by radiation, concerns that were much in the news in the mid-to-late 1980s following revelations about an excess of childhood cancers in the region surrounding a Cumbrian nuclear facility. It is of note that the connections observed in the 1950s between radiation levels and childhood cancer rates remained a matter of public concern three decades after the MRC investigation of radiation's hazards.

Kim Smith

But we can't be completely abandoned! It isn't possible![1]

Kim Smith, an English boy living in Paris, began treatment for a Wilms' tumour in the August of 1960. The first phase of treatment was to have the affected kidney removed by a French surgeon in the American Hospital, Neuilly.[2] The family were advised to submit the boy for deep X-ray treatment to tackle cancerous cells left outside the surgical area but, given the size the tumour had reached before being discovered, Kim's father was warned that the chances of recovery were not good (Smith, 1964, p. 29). The doctor in charge of the case attempted to deliver a minimal measure of hope to Kim's father, Elliott Smith:

I do not want to make things look too black. Sometimes miracles happen with children (ibid, p. 33).

Receiving radiotherapy at the Hospital Necker, a reputable Parisian children's hospital, Kim was seen by a female specialist in children's cancer (ibid, p. 46). The treatment was declared a success by the hospital staff, and normal family life resumed, with the Smith family enjoying a long sunny summer holiday in 1961. On their return, it was time for Kim to attend another of his quarterly appointments with his radiotherapist. The last two had gone wonderfully, but this time two new tumours showed up on the film: one in Kim's right lung and one on a rib, both within the field exposed to X-rays (ibid, p. 60).

Kim was prescribed further radiotherapy, this time at the French national cancer centre, the Institut Gustave-Roussy, together with a series of injections of actinomycin D to be administered at Necker. This was a relatively new drug: the earliest results showing it held promise against cancers in children had been published only two years earlier (Tan et al., 1959). The rib was also to be removed. Chemotherapy injections almost killed the child: one dose – inadvertently discharged into muscle instead of his blood stream – caused Kim's arm to break out in huge ulcers, one of which refused to heal, deepening until it perforated an artery. Kim suffered two severe haemorrhages and two emergency trips to hospital for transfusions (Smith, 1964, pp. 67, 71–5). Although the drug initially appeared to shrink the lung tumour, it quickly began to regrow at an increased rate. Further prescribed injections made Kim extremely sick each time, destroying any remaining confidence in chemotherapy his father had:

> I was now utterly sceptical about the treatment in which I had placed so much faith. It was more obvious than ever that whatever natural resistance Kim had had … was being destroyed … [I]f the cause of the cancer was not being eradicated the terrible depression of natural resistance which was in progress would be disastrous. One did not have to be a doctor to realize that. It was staring you in the face. In fact, it was advantageous *not* to be a doctor, for a layman could approach the problem with a mind untainted by prejudice or too much learning (ibid, p. 86).

Elliott Smith became convinced that treatment with radiation and chemical poisons was weakening his son's immune system and rendering his healthy tissue more prone to cancerous attack, making cure less, not more, likely. He found support for his beliefs within the medical establishment. Following the failure of Kim's chemotherapy late in 1961, Elliott wrote to the editor of *The Lancet* requesting information on actinomycin D, and was sent an article on research conducted at the Sloan Kettering Institute (SKI), which reported, as Elliott understood it, just one complete remission in 111 patients treated with the drug (ibid, pp. 80–2). Smith noted that the article also recorded tumours growing faster in those patients who did not respond as hoped. Elliott then wrote to the Chief of the Division of Clinical Chemotherapy at SKI on 13 October 1961 to ask if there were any other treatments available. The reply stated that Wilms' tumours that did not respond to radiotherapy in combination with actinomycin might remit with a combination of three chemotherapeutic agents, a form of treatment pioneered by Charlotte Tan, one of the authors of that first publication on the promise of actinomycin (ibid, p. 89).[3] Writing to Mr Smith herself, Dr Tan suggested the combination therapy could, perhaps, be administered by a French paediatric cancer specialist known to the New York group:

> The name was that of the specialist who had been treating my son, and the address was the Institut Gustave-Roussy. I did not seem to have got very far in my quest (ibid, p. 89).

Elliott sensed he was encountering a close network of doctors who would all propose treating Kim with a regime that he felt was weakening his son. In November 1961, Kim's specialist decided he needed further surgery to remove the lung tumour (ibid, p. 93). Having requested a second opinion, Elliott received a letter from Dr Tan in which she commended Kim's current doctor.[4] Elliott was not consoled:

> I could not honestly pretend to be impressed by the main contents of the letter ... I found the statement that the specialist who had treated Kim had done a 'real good job in managing his metastatic disease for such a long time' was particularly unimpressive. I have never been one to fall for facile statements, and what were the plain facts in this case? (ibid, p. 94)

Having become fluent in the language of the 'paediatric oncologists' failing to save his son, Elliott dismissed their efforts at mutual reassurance, finding no comfort in Kim's relatively long survival when compared with other patients with similar tumours. Yet Elliott could discern no substitute source of treatment to arrest his son's illness. As the interconnectedness of Paris and New York became apparent, Elliot began to actively seek out alternative approaches. His search intensified when, in the spring of 1962, recurrence of the cancer in Kim's lungs prompted his treatment to be terminated. On 24 May, Kim's specialist, having announced 'we have come to the end', refused to make further appointments (ibid, p. 102). Elliott had sought to plot his own path through available sources of help from the outset, but now, for the first time, his son's doctors were offering no options at all.

Elliott's search for alternative treatments was initially dispiriting, turning up dubious characters offering him 'secret formulas' at inflated prices, and private physicians, moved by his predicament, offering free vitamins (ibid, p. 107). By chance, he fell into conversation with a laboratory owner who advocated the use of vaccines prepared from 'microbial culture' (ibid, p. 122). Once he began to look for them, Elliott encountered many tales of 'hopeless' cases of cancer successfully treated with vaccines. The Parisian pharmacist was convinced his own product would not be enough to cure the boy, so Elliott sought stronger remedies through placing a personal advertisement in *The Times*. The editor of the paper was fearful the family would receive misleading and dangerous advice and was initially reluctant to publish the request for information. Once he had been reassured the paper would not be held responsible, the advertisement was published on 3 August 1962 (ibid, p. 129).

This was only the second time the paper had carried an appeal for assistance in finding relief from cancer. The response was immense and remarkable. Many letters recommended William Mervyn Crofton, aged 85 but nonetheless still treating cancer with vaccines from his London practice (ibid, pp. 132–3). Crofton had had a long career in pathology, publishing extensively since the 1930s on the nature of viruses and the power of immunisation to help the body fight infection. Crofton's approach to cancer had been reported in *Life* magazine the previous month. Elliott was impressed by what he learned about the elderly doctor:

> It was obvious that here was a highly qualified practitioner, not a quack selling coloured water at the country fair (ibid, p. 134).

Further letters came from patients who had received similar treatment at the Cancer Prevention and Detection Centre in London, run by Henry Augustus Morton Whitby,[5] or under the care of Roger Wyburn-Mason (Smith, 1964, pp. 135–6, 140). Wyburn-Mason believed cancer to be caused by a temperature-sensitive microbe. This man was to become Kim's final specialist.

On 11 August 1962, Kim arrived in the children's ward of Hounslow Hospital as a private patient (ibid, pp. 147–8). Given the size and number of Kim's tumours, Wyburn-Mason advised Elliott that Kim was likely near the end of his battle with cancer. Nonetheless, when pressed by Elliott, the doctor agreed to let Kim be the first child to try his new medicinal preparation against the suspected microbial cause of cancer; treatment with the supposed vaccine began the same day. The progress of the cancer was not abated, however, and Kim bled to death on 18 August (ibid, p. 168).

Later that month, Elliott Smith met with Dr Crofton, who guided him through a reading list of medical papers from the previous 50 years, which seemingly demonstrated the microbial origin of cancer (ibid, pp. 170–1). In September, the *British Journal of Cancer* published Denis Burkitt's paper about his tumour safari, in which he described how the 'African lymphoma' was restricted to areas where the temperature remained above 60 degrees Fahrenheit, implying an infectious element, perhaps borne by mosquitoes: further proof for Elliott that cancer was caused by microbes and that treatments which damaged the immune system only hastened death.

Elliott was a persistent writer and his 1964 memoir of Kim's battle with cancer, *To The Bitter End*, records a long series of consultations with numerous varieties of 'cancer experts' that continued after Kim's death. Elliott saw the researchers now regarded as pioneers of a new science as blind fools, failing to heed horrendous side-effects and poor trial results. His memoir illuminates the existence of an alternative school of thought – amongst patients' families and medical practitioners – which viewed the experimental chemical therapies for childhood cancer in the early 1960s as dangerous and counterproductive, damaging the very special circumstances that had previously permitted miracles in children: their young and energetically healthy cells and tissues. Elliott believed that Kim might have been strong enough to benefit from the vaccines if they had been administered before any radiotherapy or chemotherapy had damaged the boy's immune system. We can be close to certain that Kim would have died sooner under such a treatment plan, but Elliott's belief demonstrates the need to defend hope in the face of very poor odds: his father could not accept that Kim *never* had much chance of surviving the tumour.

Kim's story also reminds us that those children in whom the disease process was altered favourably by chemotherapy were still, at that time, very much in the minority, vastly outnumbered by those who received drugs with no benefit, and rendered almost infinitesimal by the many children with malignancies who received no adjuvant chemotherapy at all. At the time of Kim's treatment, only around one in ten children with metastatic Wilms' tumours were deriving significant benefit from chemotherapy, as measured by extra months of life; two-thirds were gaining nothing by taking part in the trials of chemotherapy drugs (Sutow, 1965). Kim's cutting-edge treatment in Paris was probably the reason why he lived so long, despite his tumour being very advanced before diagnosis, but his long painful fight against the cancer foregrounds the suffering of those families who took part in chemotherapy trials apparently – personally – for nothing. The benefits, the improved outcomes, fell to families whose children were diagnosed and treated later, with treatment protocols modified in light of the negative results seen in earlier patients. Publications in the early and mid-1960s detailing trial results of chemotherapy applied to solid tumours in children made much of the small number of otherwise hopeless cases that had been saved, and more of the optimism that some childhood tumours would become routinely curable through chemotherapy as a result of these early tests.

'Case Three'

By the early 1970s a number of British paediatric oncologists had initiated small-scale chemotherapy trials for some childhood solid tumours, employing or adapting protocols developed in the USA. As we discussed in Chapter 6, St Bartholomew's Hospital in London was one centre where such trials were managed, under the charge of Jim Malpas. Few patients were referred to the hospital from outside the London region, however, meaning recruitment to these trials was slow. Malpas decided that clinical trials should use historical controls rather than a control arm in order that the maximum number of children could be placed in the new, experimental arm of the protocol; experimental results would be compared against outcomes achieved earlier with the previous best known treatment. One trial attempted to replicate the dramatic success of two American research groups using chemotherapy for children with rhabdomyosarcoma in parallel with radiotherapy, and delaying surgery until the risk of metastases had been reduced (Figure 7.1) (Malpas et al., 1976).[6]

In this particular trial, Malpas and his colleagues treated a series of 11 children with localised rhabdomyosarcoma, administering surgery and intensive radiotherapy and a year of treatment with three anti-cancer drugs in combination. The transformation in outcome was considerable: while only two of the 17 historical control patients had lived more than 18 months, of the 11 in the trial eight were well and

Figure 7.1 Jim Malpas, Sir Eric Scowen, and Sister Fowler with a patient in the paediatric oncology unit at St Bartholomew's Hospital. Sir Eric was visiting as chairman of the Imperial Cancer Research Fund, which funded the unit
Source: Courtesy of Professor Jim Malpas, by permission of St Bartholomew's Hospital Archives.

in remission – free from evidence of disease – three years after the trial's start. The authors declared 'The natural history of this malignancy has been altered' (ibid, p. 247).[7] Claiming this remarkable advance had been achieved without causing any significant 'toxicity', they urged others to follow their example and put patients through the rigours of chemotherapy, confident the discomforts inflicted would be worth the extension in survival. The authors noted that there was still 'scepticism' in Great Britain about the value of adding chemotherapy to the treatment of paediatric tumours, so they emphasised the vast improvement that could be achieved in two-year survival for this cancer type, from five per cent with surgery alone to over 70 per cent when all three modalities of treatment were combined (ibid, p. 247).[8] They had reason to believe, therefore, that the children taking part in the trial stood a good chance of enjoying many extra months of life free from cancer symptoms, and that some might, in time, be declared cured.

The child whose experience we recreate here was 'Case Three' of this historic trial: a five-year-old girl treated for rhabdomyosarcoma of the eye. Case Three had one of the most positive outcomes of the trial: at the time the report was published she had remained disease-free for 29 months. The trial report provided considerable detail about specific patients, showing variation between children, as well as picking out common responses; the publication was issued as an indication of early promise in a research area, intended as a rallying call to other researchers to work in cooperation, answering questions over optimal drug doses and timings, and exploring how the results might be applied to more common tumours in adults (ibid, p. 249). We can reconstruct much of the treatment for Case Three from these details, but not her emotional state or her level of family support. While this account offers a pale sketch when compared with the pictures painted in patient or parent memoirs, it nonetheless holds value for what it reveals of clinicians' awareness of the impact of strenuous treatments on their patients, and for the window it offers into clinicians' attempts to understand whether treatments might be 'too much' to bear.

Case Three's tumour was deemed to be regional and resectable: it extended outside the eye itself but could be removed surgically, and she had no visible metastases. Following an operation to remove as much of the eye tumour as possible, she immediately underwent a five-week course of radiotherapy, receiving 1,000 rads per week, each week's dose being split over five days. She would probably not have been treated as an inpatient, and therefore would have been travelling to the hospital frequently.[9] Chemotherapy was started at the same time as radiotherapy, and consisted of three drugs administered concurrently: vincristine (two patients could not cope with its effects, and were therefore given an alternative drug), actinomycin D, and cyclophosphamide. Six cycles were administered, and if Case Three's blood count remained high enough these would have been spaced at weekly intervals. If her blood counts dropped, treatment would have been paused to give the bone marrow a chance to recover from the suppressive effects of the drugs and radiation. The authors noted that interruptions to the chemotherapy routine were common and frequent.[10]

During the first six to eight weeks of treatment, the girl would have made five visits to the hospital each week, taken weekly blood tests and injections, and experienced reddening skin, nausea, and perhaps some loss of sensation in the fingers and toes. After a few weeks she would also have become anaemic, rather tired, and susceptible to infection and bruising, and she would have lost her hair. After the first two months, her visits would have become less frequent, with chemotherapy administered fortnightly until a full year of treatment had been completed. We also know that Case Three underwent a second operation after eight months to remove her eye entirely, during which no remaining tumour was found (ibid, p. 249).[11]

So in what respect can this treatment be described as well tolerated and as being without significant toxicity? This is where it is impossible to move from the clinician's standpoint to recover the patient's, for we have no way of knowing what this five-year-old made of her new routine, her side-effects, and the loss of an eye. Was she psychologically scarred by these experiences, or did she take them in her stride? Did she have a family who was able to make her still feel lovable, and tolerably comfortable, emotionally if not physically, through the rigours of her treatment, or did her family visibly suffer throughout the process, leading her to consider her illness a burden on them all? We cannot determine whether this was simply a longer remission than previously achievable, or a full cure, without access to further long-term follow-up data on the recruited patients, and we do not know how well informed the family were before agreeing to take part in this clinical trial. What is clear, however, is that as with Kim Smith, her participation in a clinical trial helped further understanding of the power and limits of chemotherapy to improve the likely outcomes for patients. She was also given the 'best known' treatment at the time, as was understood by the most informed paediatric oncologist of the day. Even so, her long-term prospects were uncertain: as the addition of chemotherapy to treatment regimes started to yield dramatic jumps in survival times, no one could know whether long survivals would turn out to be cures, or what to advise families about their futures.

To Tell or not to Tell

One of the issues raised by these two accounts is that of disclosure: whether a doctor should inform the child with cancer, or their family, of their diagnosis and prognosis, let alone of what their likely treatment options are, and what risks and side-effects these might bring. Elliott Smith was told his son's chances of complete recovery were poor, but was not apparently given any percentages on which to base his decisions about treatment options until Kim's French specialist (who it can be deduced must have been Odile Schweisguth) announced that death was certain. We do not know what Jim Malpas told Case Three or her parents about the future they faced: a chance of one in 20 for a two-year remission without the additional chemotherapy, and an improved but hard-to-predict chance with the experimental treatment. But even apparently 'hard' statistics like these were little use when each new trial result suggested very different chances: the experiences of past patients could not be taken to be an indication of the odds faced by those newly diagnosed.

In the USA and in Great Britain, medical teams debated during the 1960s how they could best help patients – children or adults – and their families prepare for cancer treatment; a need for honesty was preached by some, but truth-telling was not widely practised. Even into the 1970s, there was little consensus about the issue for child patients or for adult cancer sufferers, with many commentators believing knowledge of the truth would dispel all hope, and blight the patient's remaining time.[12] The situation began to change when, rather than focusing on what knowledge would benefit or harm the child patient, observers instead began to question what they might already know. Ruth Frank Baer, a former Instructor of Pediatric Nursing at the University of Illinois, addressed this issue in her contribution to a 1955 book, *Should the Patient Know the Truth? A Response of Physicians, Nurses, Clergymen, and Lawyers* (Standard and Nathan, 1955). Following a discussion on the subject at a meeting of the New York Academy of Medicine, it was agreed a book should be commissioned, with the contributions to be as wide-ranging and inclusive as possible. Although the book was not specifically addressing any one disease, cancer was by far the dominant theme.

On the issue of informing parents that their child was dying, Baer had conflicting views:

> I believe parents want to hear the truth about their sick child. They may not accept it when told, but they would become angry with any doctor who lied to them or withheld information. As a parent I feel strongly about this, though I am not so sure I would always want to know the truth about myself … To a nurse caring for another's child, the other very different side of the coin presents itself. I have often felt that it is not fair or right for any mother to know her child has leukaemia. Why must she know for so long a time that her child will die? Why must she mourn the child with each exacerbation and have false hopes with each remission (1955, p. 104).

Despite supporting the general pattern of not telling a dying child the truth, Baer believed that many were well aware of their situation:

> We do not tell, but somehow children seem to know, like Jerry, age eight, who would look at you wisely and say, 'Soon I'm going to fly

right out of here.' And we, not wanting to accept the truth ourselves, think Jerry has been reading Superman too much. No, we do not tell the truth, for with children it is difficult to manage it ourselves (ibid, pp. 104–5).

Audrey Evans, a respected American oncologist who developed the Evans Staging System for neuroblastoma, published an influential paper, 'If a Child Must Die', in 1968. Evans' starting point was a description of existing practices, characterising the two opposing camps' conflicting opinions as to what was in the best interests of the child:

> There appear to be two basic differences of approach to the care of the dying child. One emphasizes helping the patient to face death and the fears involved with dying; the other attempts to shield him from death and protect him as much as possible from such fears (1968, p. 138).

Although Evans believed either method might be correct 'in the right hands', she favoured shielding children from the full knowledge of what was to come, a doctrine she traced back to the practices of Sydney Farber. Born in York, England, Evans had studied at the Royal College of Surgeons in Edinburgh, before travelling to the USA in 1953 as a Fulbright Fellow. She trained under Farber at the Boston Children's Hospital, where he established his 'total care' movement, within which the families of leukaemic children were treated as a whole, their psychological, social, and economic needs being addressed, as well as their medical needs. So inspired, Evans would go on to co-found the first Ronald McDonald House for the temporary accommodation of families of child cancer patients in 1974, while serving as Chair of Oncology at the Children's Hospital of Philadelphia.

Evans recommended disinformation, telling children with leukaemia they had anaemia, and children with solid tumours that their symptoms were the result of abnormal cells building up like an abscess. She also cautioned against parents trying to find out more information about their child's illness, on the grounds that anything they located would most probably be out of date, and therefore might depict the disease as hopeless or, conversely, be 'overly optimistic'. Either of these could damage parents' ability to both cope with their child's illness and relate sufficiently strongly to the physician in charge (ibid, p. 139).

Her advice extended to lying outright to a child even during their last days, in order to spare him or her from anticipating the horror to come. Hope that a last minute miracle might occur was to be offered to child and parent alike, as the loss of hope was seen as particularly destructive. Evans believed the majority of children did not consider death a possibility:

> One is rarely asked directly by a child if he is dying; small children have little concept of dying as a result of illness and in relation to themselves, but occasionally the fear of dying is voiced by a teenager. I usually concede that he is very sick, and that I can understand his anxiety, but say that I do not know when anyone is going to die. I remind him that we have always been able to make him better, and I have known children who were as sick as he who recovered. Sometimes, anxieties about the future are hidden in such casual questions as going to Junior Prom in June or learning to scuba dive next summer. The answers should be geared accordingly and can be helpful, but indefinite (ibid, p. 141).

Evans was not alone in believing such manipulations of the facts were of comfort to patients and their families: the majority of medical care teams looking after the sickest children with cancer in the late 1960s honestly believed it was better for dying children not to know they were facing death.

Myra Bluebond-Langner shifted the terms of this debate, having spent time with leukaemia patients in a children's ward to examine what they already knew. Her study was undertaken as research for a doctoral dissertation in Anthropology at the University of Illinois, and was carried out over nine months in the Department of Pediatrics of a large, midwestern hospital in 1972. The thesis was published six years later as *The Private Worlds of Dying Children*. This was the last major study of children with cancer to assume and find leukaemia to be universally fatal: all 32 major informants died prior to publication (Bluebond-Langner, 1972, p. x). Bluebond-Langner's subjects all followed a similar path of decreasing engagement with daily life as they became certain they were not going to get well.

Bluebond-Langner sought to persuade her readers that children with leukaemia knew much more about their condition and its likely end than was generally accepted. By spending many hours both observing and conversing with each child in many settings, including, when possible, their homes, and witnessing their interactions with family members, staff and visitors, she developed a powerful critique of the widespread assumption that it was better not to tell children what was making them unwell or what lay ahead. She demonstrated that while children might not ask questions about their health and what the future may hold for them, this was not due to a lack of interest. Rather it was a sign of their socialisation and awareness of social order, 'of how it is acceptable to die in this society' (ibid, p. ix):

> Since the children interpret death as an inappropriate topic of conversation with adults (evidenced by the adults' reactions when children try to discuss it), and as an appropriate topic with other children (evidenced by their willingness to offer information and answer questions), they refrain from discussing the subject in the presence of adults, but pursue it with peers. This is also true of sex discussions among normal children. Leukemic children often discussed their condition in the place children often go to discuss sex – the bathroom, where adults cannot hear them (ibid, p. 10).

Bluebond-Langner characterised parents as frustrated nurturers, and doctors as impotent healers, each unable to fulfil their proper roles in the face of an unstoppable disease process; wishing to preserve the feelings of their parents and doctors, children feigned ignorance of their fate in order to enable carers to feel less helpless. She observed that normal life, both at home and in hospital, could only be preserved through a sustained 'mutual pretense', a shared lie that the child had a future.

Bluebond-Langner noted that as children moved through successive remissions and relapses, the future ceased to have meaning for them. They lost their initial courage and optimism once in their first relapse and, having witnessed those around them die, became convinced this was to be their fate as well:

> The children's view of time itself changed … It was no longer, as it is for most other children, endless. It became finite, marked by relapses and remissions. One consequence of this changed view was that the children no longer spoke of the future. Above all, time was not to be wasted (ibid, p. 12).

There were few grounds, even in 1972, to hold onto hope that any one child would be the lucky one, despite this year being hailed by some medical historians as the point at which acute leukaemia could be said to be cured (see, e.g., Le Fanu, 1999). Bluebond-Langner's personal reflections on the project recount her own disillusionment:

> I think of all the progress in cancer research, of the advertisements from research centers showing 'cured' children ... [I]t is said that some children have already been cured, but of what have they been cured, and for how long? (1978, p. 253)

Her subjects and their fate made Bluebond-Langner feel both angry and guilty because the children she studied would not *become* anything (ibid, p. 254). Cured children were too few and far between in the world of ill children she studied to give the concept of 'remission' any curative connotation. 'Remission' for her, as for the children she studied, meant only a temporary reprieve on the road towards death.

By the time *The Private Worlds of Dying Children* was published, however, an alternative challenge had arisen within paediatric oncology. As treatment success rates rose for both leukaemia and the previously 'hopeless' cancers, physicians and the experts in psychology they consulted were observing a new problem. Increasing numbers of children with cancer were living beyond their period of treatment and returning to their 'normal' lives, albeit lives often blighted by the long-term consequences of that treatment, and by uncertainty. They and their parents needed advice on how to manage the long period of uncertainty.

Dealing With Uncertainty

In the 1970s, America was home to an increasing number of previously unanticipated survivors of childhood leukaemia and other cancers. Specialist conferences were organised, usually with patient and parent participants, solely to discuss the ongoing needs of the cured. The first large meeting dedicated to the problems of cured children and adults was held at the Georgetown Medical Center, Washington DC in 1974, with the meeting's report taking over an issue of *Cancer*, the prestigious journal of the American Cancer Society, the following year (*Cancer*, 1975a, 1975b). Those present took the psychological sphere to be the location of the most severe problems for the cured. With family, friends, insurance companies, and employers often unwilling to believe that cancer was curable, survivors were struggling to adjust to a culture in which cancer was seen as permanently life altering.

In Great Britain, no meetings on the problems of the cured were held until the early 1980s. In the mid-to-late 1970s, however, the largest treatment centres developed links with child psychiatrists and psychologists in order to offer support to families and gain insights into making treatment less psychologically costly. British studies were often modelled on work undertaken in America, and were cited, in turn, by American studies; since the mid-1970s the two countries have consciously shared one body of literature, those working in the field taking studies conducted in either country to apply to both. The research that came out of this interest in understanding the needs of the potentially curable, and supporting the maintenance or re-establishment of normal life after treatment, came to be known as psycho-oncology. An early investigation into the social and psychological aspects of childhood cancer was undertaken in Great Britain. Its premise was wholly different from that taken by Evans and Bluebond-Langner; the question addressed was not how to prepare families for (almost) certain death, but how to help families handle not knowing what the future would bring.

Between January 1976 and April 1977, Jean Comaroff, an anthropologist from the University of Chicago, and Peter Maguire, a psychiatrist at the University of Manchester, studied 60 families with children suffering from acute leukaemia being treated in the wards and clinics under the charge of Pat Morris Jones at Pendlebury, Manchester. The resulting paper presented the existing orthodox view amongst British paediatric oncologists regarding the proper way of supporting parents. Comaroff and Maguire observed that British paediatricians, in order to give families time to prepare, found it preferable to be frank once death was certain. However, there was no established position on how to support families facing the possibility of extended life (Comaroff and Maguire, 1981, p. 123).

The families surveyed experienced childhood leukaemia as a highly uncertain disease: not only uncertain in origin and outcome, but also unstable medically, subject to the assertion and retraction of new knowledge claims on a regular basis (ibid, p. 115). The meaning and significance of remission had changed markedly since the early 1970s, with good outcomes increasing and published estimates of cure rates changing the stated chances of life frequently. This fluidity in prognosis appears to have made the problems of coping with uncertainty even more acute:

> the experience of uncertainty and the search for meaning were *the* characteristic features of the impact of this disease upon sufferers and their families (ibid, p. 115, original emphasis).

Comaroff and Maguire suggested the lack of certain medical knowledge about leukaemia made uncertainty over outcome even more threatening than it would have been had medical knowledge been secure. Living on the medical frontier exacerbated patients' and parents' psychiatric distress. The only amelioration was mixing with other families in the same situation: families used one another to construct new norms against which to measure their own progress.

Comaroff and Maguire observed that clinicians typically promoted an attitude of living in the present, encouraging children and parents to consider only the immediate future. The families in their study, however, attempting to conduct normal lives, expressed a desperate need for a longer-range view. One issue raised was whether to discipline badly behaved children in order to socialise them ready for adult life, with one mother remarking: 'If he's not going to grow up, what does it matter?' Dwelling only on the state of remission could not guide proper family functioning, mere survival being of no comfort without some indication as to its meaning, its prognostic value for the future. The experience of leukaemia had no moment of resolution, only an endless waiting to see if a child had indeed escaped death.

Comaroff and Maguire focused on the experience of uncertainty for what it revealed about the meaning of the disease in general, for clinicians and for lay people, and for the window it afforded on patients' perceptions of 'the meaning and value of biomedical science' itself. Such a critique of medical science, drawing on the work of Michel Foucault, was cutting-edge in the sociology of medicine at the time. They drew attention to the totemic value of cancer, which they described as having 'become a standardised nightmare in our society' symbolising the 'ambiguities of technical control', the ambiguity being that the more medicine appeared able to control, the more frightening became those diseases still beyond its powers (ibid, p. 116). They also stressed the etymological difficulties elicited by the use of the word 'remission':

> [the term] ... entails a range of meanings which combine the notion of divine pardon with the retreat of symptoms. It represents in condensed form the entanglement of control and chaos at the frontiers of medical knowledge (ibid, p. 120).

This study into patients' and parents' experiences of survival, while one of the earliest, was far from unique in concluding that coping with uncertainty and living in a permanent present were the principal issues for families with leukaemic children; in the concluding chapter we explore more contributions to the burgeoning field of survivor studies. Here, we wish to look in detail at one family's experiences of living with uncertainty, and with the chance and randomness that permeate clinical trials, drawn from the same hospital in which Comaroff and Maguire's milestone study took place.

Jimmy Renouf

> One of the first facts to grasp is that each child, in order to be treated at all in this country, has to be a guinea pig (Renouf, 1993, p. 40).

Jimmy Renouf was living in a Lake District village with his parents and younger brother, Martin, when he was diagnosed with acute lymphoblastic leukaemia in 1986. He died in 1990 at the age of eight, and a book about the family's experience written by his mother, Jane, was published in 1993. Her account gives considered insight into the lived experience of taking part in a clinical trial where one cannot know whether the trial arm to which one is randomised is going to deliver better or worse results than the other treatment options under test.

Upon his diagnosis, Jimmy and his mother were sent to the Royal Manchester Children's Hospital, Pendlebury, where they stayed for just over a month. Jimmy's details, like those of every patient with the condition, were fed into the United Kingdom Acute Lymphoblastic Leukaemia computer, which then allocated each child randomly to one of four protocols. Results were then examined every four years and protocols modified depending on their success and failure. Despite Jimmy's white cell count being extremely high, he was allocated protocol A, the least powerful of the treatment options, with no intensification of treatment after the first month or repeats of the initial treatment as in protocols B, C, and D. Jane Renouf eloquently addressed the pros and cons of this system:

> Just supposing there were a choice, and we were to select a level of treatment which ultimately failed, would we not blame ourselves forever? And if the doctors chose for us, would we not blame them? And if the computer selection system contributes to the current research, without which no improvements can be made for the future, wasn't this really the only sensible way a decision could be made?
>
> It is only with hindsight that I think that the computer decision to give Jimmy as little treatment as possible was wrong; and that the morality of random selection is questionable, however much it

lets everyone off the hook by taking a decision nobody wants to be responsible for (ibid, p. 41).

Not long after Jimmy's death, this random selection procedure was modified, giving more consideration to white cell counts when selecting treatment protocols.

Unfortunately, Jimmy was one of only around five per cent of children who relapse while still undergoing chemotherapy. The options available were to do nothing, the prognosis being death in eight to ten weeks, or to recommence more intense treatment, in which case Jimmy would also require a bone marrow transplant. Even if remission was achieved and a perfect donor match found, there was only around a 20 per cent chance of cure. Jimmy himself made the decision to fight on.

Although further chemotherapy proved unsuccessful, the family were offered daily chemotherapy tablets in the hope they could keep Jimmy alive over the Christmas of 1989. In the New Year, a thumb-prick blood test revealed cancerous blast cells once again circulating in his bloodstream. Jimmy was also becoming increasingly anaemic and was given a blood transfusion, something he did not initially want:

> Perhaps we should have recognized his greater wisdom. He grudgingly agreed to have a transfusion, which, it transpired, may well have been responsible for prolonging his life beyond quality and into pain during the next few weeks (ibid, pp. 138–9).

When the pain started, Jimmy's local general practitioner (GP) visited every day, administering morphine to little effect. The family felt abandoned by the staff at Pendlebury, having received insufficient information about what to expect, or what equipment – such as a battery-operated syringe driver – they might need to ease his final, painful weeks. Several times, Jimmy wished he was dead:

> How much better for Jimmy if he had died that first weekend of pain. He had prepared himself, told us the important things he wanted us to know, settled his affairs on earth and looked forward to his journey to Heaven (ibid, p. 153).

Having barely been expected to see Christmas 1989, Jimmy eventually died a fortnight before the following Easter. Jimmy was thus one of the 26 per cent of British leukaemia patients diagnosed in the period 1986–90 who did not survive to the five-year point (Stiller, 2007, p. 176).

In her memoir, Jane Renouf questioned whether there was a mind/body connection in relation to cancer. Jimmy never forgave his brother Martin, two years his junior, for being healthy:

> We longed for reconciliation between the two, and, in the absence of any definitive medical explanation for his illness, we even began to wonder if the depths of Jimmy's anger weren't partly to blame for his leukaemia returning. Could it really be true that an unhappy mind caused a sick body? Had Jimmy been unhappy all his life? (1993, p. 130)

Echoing beliefs about the relation between cancer and the immune system that we have seen others express, Jimmy's parents wondered if hypnotherapy might 'reactivate' his body, inspiring it to once more fight the cancer. The hope this instilled gave them 'a short break from despair', the local GP's wife conducting sessions with the boy she had known his entire life. Although the techniques she taught him to control the pain he would later suffer were limited in their effectiveness, Jimmy enjoyed the sessions, particularly the stories intended to make him feel calm, happy, and loved (ibid, pp. 130–1).

As Jimmy's story demonstrates, clinical trials, widely acknowledged as one of the greatest breakthroughs in medical history, can appear very different to those on the inside. All the evidence available – memoirs, correspondence, and interviews – suggest the meaning of 'patienthood' is transformed once patients are enrolled into scientific experiments: patients become data points, as well as sufferers. While patients as ill people can produce meaningful stories by virtue of their characters and relationships, patients as participants in experimental trials also generate meaningful scientific data, leading to aspects of their experiences being abstracted as case reports informing clinical experimentation. It is helpful that some patients and families produced rich accounts of their efforts to make sense of the multiple meanings of cancer and of being a patient during this period of clinical experimentation. Doctors' recollections, too, whether in print or accessed through interviews, reveal their awareness of how clinical trials added an extra layer of meaning to being a patient.

Fears of Radiation

Jane Renouf speculated in her memoir about the origins of Jimmy's original piece of bad luck: contracting leukaemia in the first place.

She wrote at some length about the family's close proximity to the nuclear facilities at Sellafield, the West Cumbrian nuclear plant formerly known as Windscale, renamed in 1981. She related how Jimmy and his brother were taken to the beaches along the Cumbrian coast to swim in the sea during 1983, a year when it was found that high levels of radioactive ruthenium and rhodium had been discharged into the sea and that ended with the imposition of a 12-month closure of the beaches along that stretch of West Cumbria.[13] Although the family had been aware of a 1977 inquiry into the feasibility and safety of reprocessing being carried out at Windscale, Jane reported they were not originally overly concerned about living close to the plant.[14] Gradually, however, once Jimmy had been diagnosed, they became aware of other local children suffering from cancer, particularly leukaemia. Having realised there had been five cases of cancer in the past four years within a child population of less than 1,000, they wrote to their local member of parliament, Michael Jopling, who informed the then Minister for Health, Kenneth Clarke. Following an exchange of letters, the family were informed the children did not constitute a 'cluster' of cases, a circumstance that might have suggested a local cause. The population was deemed too small and the cases too few to be 'statistically accountable' (Renouf, 1993, p. 67). Despite this, researchers in Newcastle suggested the affected children be scanned for levels of a radioactive isotope of selenium present in spent nuclear fuels and in the waste products generated during their reprocessing. The family waited for an appointment to attend Withington Hospital in Manchester for a body scan but no letter ever arrived; another family who did attend an appointment were sent home, there being no one available to perform the scan. Jimmy was placed in his coffin wearing his favourite t-shirt, depicting the Greenpeace ship, *Rainbow Warrior*, which he and other Cumbrian leukaemia sufferers had visited in London before his relapse.

The Nuclear Laundry

Three years before Jimmy was diagnosed with leukaemia, an excess of cases of cancer in the young reached the attention of the public at large owing to a televised report by a small team of investigative journalists. On 1 November 1983, Yorkshire Television (YTV) aired a documentary entitled *Windscale: The Nuclear Laundry* in which it was claimed that living close to the British Nuclear Fuels Limited (BNFL) nuclear plant on the west coast of Cumbria magnified a child's chance of dying of cancer by ten. The region appeared to have a particularly large number of incidences of leukaemia and lymphoma, with the nearby communities of Seascale and Bootle experiencing 11 and four cases, respectively, of these blood and lymph cancers over the previous 30 years. For such small populations, the expected number was only half a case. The programme suggested that plutonium, a radionuclide removed from spent uranium fuel during the reprocessing procedure, was much more detrimental to health than commonly believed. As Windscale emitted far more plutonium than other nuclear installations in Great Britain, the implication was that absorption of particles through food and their surroundings had caused the children's cancers. Plutonium was known to be a 'bone seeker', an element that the body retains and thus delivers a far higher level of radiation to bone marrow cells than suggested by the levels of radiation in the environment around the plant.

Responding to the strong and immediate media and public outcry, the government announced an official inquiry the following day. As this was being organised, individuals involved with the programme, the methodology employed, and specific claims, were challenged by various interested parties. The YTV team, having initially intended to examine the effects of industrial exposure to radiation on workers, had quickly noted the high rate of leukaemia in local children since the plant had opened, and shifted their focus, making door-step enquiries to identify possible cases of childhood cancer in nearby villages, then requesting death certificates from local registry offices to verify cause of death: not a standard practice for establishing disease incidence or excess. Epidemiological studies established to demonstrate an excess are open to the charge that the boundaries of the region or population surveyed have been decided upon to maximise the significance of data collected – akin to shooting a series of holes in a plain board and only then drawing on the target in order to maximise the number of bulls-eye hits.[15] However, this was one of only two methods available, as requests for information on cancer deaths made to local registrars by the YTV researchers had been rejected. The second method used by YTV – and subsequently by Cumbrian antinuclear campaigners – was to visit local churchyards, locate gravestones of young people, and then personally search the district death records for the cause of death in each case.[16]

The credentials of the two independent academics who had measured radiation levels, Dr Philip Day from the Victoria University of Manchester, and Professor Edward Radford from the University of Pittsburgh, together with those of the statistician John Urquhart, were also questioned.[17] Professor Radford (1983) sent an angry letter to *The*

Financial Times bemoaning the apparent lack of governmental concern about the excess of cancers along the Cumbrian coast, and compared the dismissal of the documentary's claims to the similar response to the 1977 Windscale Inquiry. This investigation into a long history of leaks and high emissions had reassuringly concluded that radiation levels were low by international standards. He claimed, however, that dangers existed at levels far lower than was admitted by those international bodies responsible for determining what counted as 'safe'. Reversing the argument, Radford declared that these standards could not possibly be safe given the excess cancers on the Cumbrian coast.

The inquiry was chaired by Sir Douglas Black, former President of the Royal College of Physicians and chair of the 1980 Black Report on health inequalities. His team included experts in epidemiology, oncology, and paediatrics.[18] Their report, 'An Investigation of the Possible Increased Incidence of Cancer in West Cumbria', was published on 23 July 1984, and a television programme, *Windscale: The Black Report*, was aired the same day. The Black Report was immediately met with derision, with sceptical journalists labelling it a 'blackwash'. *The Guardian* was the most damning of the national newspapers, drawing attention to the circular reasoning used in the report. Having quantified a local person's exposure to radiation from the plant as 13 per cent over and above the background radiation, it declared this figure could only account for an additional 0.091 deaths in the area over the time period studied, as opposed to the excess 3.5 deaths recorded. The report used this discrepancy to argue that the extra deaths could not possibly be caused by the radiation emitted. But, as *The Guardian* stated, 'the figure which counts is the number of deaths which occurred, not the number the National Radiological Protection Board (NRPB) would have expected': the onus of proof was not as Black had assumed. As far as the public was concerned, Windscale/Sellafield had to be proved innocent, and those biological models that alleged radiation could not be the sole cause had to be proved right in the face of a persistent and suggestive excess of deaths (*The Guardian*, 1984).[19]

Despite the fact the Black Report could neither explain nor discount the ten-fold increase in the risk of leukaemia or lymphoma, it still called for calm, urging local medics to reassure the public. Barrie Walker, a local GP, took vocal exception to this. Having treated several of the Seascale cases, Dr Walker said there was 'no way' he could reassure people, given the children at his practice under the age of ten had a one in 60 chance of developing a malignancy compared with a national average of around one in 600 (*The Daily Telegraph*, 1984).

The Black Report concluded that further, substantial research into the effects of radiation exposure and the routes of radiation exposure in children were the most promising lines of inquiry. The few studies of radiation's effects then available were derived from medical follow-up of children who had experienced high-dose but very short-term exposures to radiation, chiefly the Japanese survivors of the atomic bombs discussed in Chapter 2. Whether these children could serve as a valid comparison group to the West Cumbrian children was not discussed. However, the pattern and type of exposure were completely different, and one can assume that many of the exposed children in Hiroshima and Nagasaki died well before any radiation-induced malignancy could be detected. The report recommended a series of animal studies to monitor the prenatal absorption of radionuclides through the placenta and, during infancy, absorption via the gut lining, and to quantify any additional effects to marrow caused by the relative smallness of the bones and the fact they were developing throughout exposure. More controversial were suggestions to the Ministry of Health that experiments be initiated on the local population to measure the actual radiation exposure of different groups of people in the vicinity of Sellafield, suggestions that were missing from the published report.[20]

Dissatisfaction with the government's response to the YTV programme and the questions surrounding nuclear safety did not dissipate quickly. Five months later, in his article 'Why Radioactive Doubts Remain', Robin Russell Jones bluntly summed up widely held concerns:

> The Black Report is not a scientific document. It is an exercise in public reassurance which has gone disastrously wrong for the simple reason that there is no scientific basis for its conclusions (*The Guardian*, 1984).

Further embarrassment was caused in 1986. Dr Derek Jakeman, a former employee of the UK Atomic Energy Authority, having found no mention in the Black Report of the substantial uranium leaks he had monitored between 1952 and 1955 while the plant was still a military site, went public with the fact that radioactive emissions in the 1950s had been 50 times higher than previously admitted.[21]

An alternative hypothesis for the excess of cancer cases was the possibility of a viral cause. As we discussed in earlier chapters, the idea that cancer might be caused by a microorganism has a long history, and despite the low number of confirmed relationships between viruses and specific cancers at the time of the Windscale controversy, the theory remained attractive. Were a viral cause to be discovered, the advantages for the government and the nuclear industry would be obvious, but the prospect of a viral cause also held appeal for the wider public, offering hopes that medical science might isolate and respond to the health threat.

The Black Report considered possible bacterial or viral sources for the excess of cancers but concluded that viruses with carcinogenic potential in humans were not relevant in this situation.[22] Interest in a possible viral cause for the excess nonetheless persisted. In 1989, Richard Doll, Great Britain's most influential epidemiologist at the time, wrote a letter on the subject to another Cumbrian mother, acknowledging there were several types of cancer known to be caused, at least in part, by viruses, but stating that he was 'far from convinced' that a virus had played any part in these cases. Only one factor, he noted, had been firmly established as a cause of childhood leukaemia, and that was ionising radiation. Doll then dismissed the possibility that emissions from Windscale could explain the Seascale excess, not only because the doses received by people in the area were much smaller than those experienced from background radiation by everyone in the country – a contentious claim, as we have seen above, once one considers the pathways by which plutonium might have reached the bones of children – but also because 'the excess is limited to leukaemia and radiation is known to cause a proportionally equal increase in all types of childhood cancer and no general excess of other cancers has occurred in the area'.[23] Doll was referring to the findings of Alice Stewart, and his use of her results is curious, given that at the time of her findings he opposed her claim that radiation could be responsible for increases in all childhood cancers and not just acute leukaemia.[24] His justification then had been that the available biological models did not offer a mechanism for radiation to increase the risk of solid tumours and that the data from Japanese bomb survivors revealed excess cases of leukaemia only. Thirty years on, Doll's opposition to Stewart's assertion that radiation increased the risk of *all* childhood cancers had disappeared: instead, that finding was being deployed to defend the radioactive outputs of Windscale/Sellafield against the charge that they were responsible for the excess deaths observed.

Although debates continue in radiobiology journals and conferences, the question of whether radiation can cause leukaemia and other cancers in the young no longer excites the public imagination to the same degree. The strong and unique linkage in the public mind between ionising radiation and cancer in children appears to have been broken, despite the epidemiological observations in the wake of the Chernobyl disaster and the more recent problems at Fukushima. Attempts to bring the science of carcinogenesis to the public have diminished: it is deemed too complicated or too uncertain to excite the same dedication and commitment in journalists; therefore, the scientific basis of stories rarely makes it into mainstream news. Nonetheless, since the early 1980s one of the available cultural meanings of childhood cancer has been that the sufferer has fallen prey to mankind's pollution of the planet with ionising radiation. Environmental campaigns in many countries call for clean-up efforts and compensation wherever excesses of cancer in the young are observed, whether these be proximate to chemical plants, by concentrations of electrical pylons, or near waste dumps. The meaning of childhood cancer can be shaded by the uncertainty of possible causes, as well as the uncertainty of prognostic chances on different clinical trial protocols and of ultimate treatment outcomes.

8
Experiences of Survivorship

In the previous chapter, we explored the recorded experiences of three children who received experimental treatment for their cancers, in the early 1960s, the early 1970s, and the late 1980s, respectively, and we examined influential sociological studies on the practical and psychological problems faced by families living with the uncertainty of remission, intended to help medical professionals provide useful information. We noted that by the mid-1970s, increasingly successful treatments for childhood cancer prompted both paediatric oncologists and psychologists to focus on minimising the burdens carried by those patients surviving into adulthood and their families. In this chapter we consider the emergence of the field of psycho-oncology and its continuing importance, for medical professionals and cancer survivors alike.

In 1981, the year Comaroff and Maguire's exploration of the changing meaning of childhood leukaemia had emphasised the psychological distress resulting from the uncertainty of medical knowledge and instability of remission, the results of another hugely significant study were also published. This research, specifically focused on the ways in which long-term survivors of various childhood cancers adapted to their situation, would influence all future work in the field. Appropriately, exemplifying as it does the second tenet of Farber's (1965) concept of 'total care' – the need to exert 'all possible efforts to increase and maintain the happiness of the patient and the mental peace of the family' – it was carried out at the Sydney Farber Cancer Institute in Boston. Between 1976 and 1981, a team led by Gerald Koocher and John O'Malley held in-depth interviews with 117 survivors, assessing their physical and mental health, levels of anxiety, social adjustment, and self-esteem. Koocher and O'Malley (1981) named the condition survivors reported finding themselves in – one of perpetual uncertain hope – after the ancient Syracuse courtier invited to a lavish royal banquet, but forced to sit under a sword suspended by a single thread: *The Damocles Syndrome*.

Koocher and O'Malley's main focus was whether patients were able to perceive themselves as cured and 'move on', leaving their cancer experience in the past and, if so, when this could be achieved. While outlining their methodology and its relation to earlier investigations, the authors explicitly stated that, unlike many earlier researchers, they would not pathologise any of the diverse coping styles they observed (Comaroff and Maguire, 1981, p. 116).[1]

Defining Coping

Studies of strategies for coping with cancer have tended to focus on adults, and on breast cancer patients in particular. From the mid-to-late 1970s, researchers led by H. Stephen Greer at The Faith Courtauld Unit, King's College Hospital Medical School, had claimed that breast cancer patients adopting a 'fighting spirit' were less likely to suffer a recurrence than those that respond with helplessness and hopelessness (see, e.g., Greer et al., 1979; Greer, 1981; Pettingale, 1984; Pettingale et al., 1985). The idea of cancer as a metaphorical war, requiring patients to always 'think positive' and 'fight' the disease – in a way not expected when dealing with other illnesses – has been persistent, but has been artfully deconstructed by many researchers.[2] Although few studies pathologising coping styles have related to children, the military metaphor and the dogma of positive thinking in relation to cancer have become practically ubiquitous and enduring. At a 2004 multidisciplinary forum on 'Cancer as Metaphor' at Massachusetts General Hospital, a paediatric oncologist described a six-year-old girl who continually lashed out at medical personnel attempting procedures. Eventually, a nurse realised the child had misunderstood her mother's constant urging to 'keep fighting', to 'beat the cancer' (Penson et al., 2004). The possible consequences of such metaphoric conceptions of cancer for young children have been addressed. Since the publication of Lynn Baker's 1975, *You and Leukemia*, by the US National Cancer Institute, information for very young cancer patients has tended to depict the disease as the result of confused rather than malevolent cells. Unsure of what they were supposed to be doing, bewildered cells would crowd together for company and form tumours. By presenting cancer as muddled rather than warlike, it was hoped young children would not feel too badly let down by their bodies, and therefore avoid the alienating process of 'splitting-off', which could result in self-hatred (Barnes, 2006).

A need to stop pathologising the coping strategies employed by families of childhood cancer survivors had been remarked upon in psycho-oncology literature prior to *The Damocles Syndrome*.[3] In one particularly thorough review of studies conducted in the late 1970s, the authors declared such a change necessary, not only because possible outcomes had been so altered by medical advances, but also because there was no longer any predetermined ideal path to acceptance of prognosis. The new aim for parents and patients was to manage a life in limbo, as it were; thus, researchers needed to revise their value judgements about coping strategies that emphasised optimism in the face of modest chances (Koocher and O'Malley, 1981, pp. 29–30). In this new school, dated to approximately 1977 but with antecedents in studies of the early 1970s, even denial was to be respected, its value recognised (ibid, p. xiv). This has been explained as a consequence of general moves within psychology intended to establish the discipline as medically *useful*. If measures could be devised to identify which coping strategies were more likely to lead to well-adjusted survivors and families, then the role of the psychologist within the care team would be assured.

Clinical trials to rate the usefulness of various psychological interventions were conducted in the USA. In Great Britain, however, the majority of studies were designed to map the coping strategies already employed by families of childhood cancer patients, with the expectation that such understanding might help clinicians predict which families could benefit from intervention, and of what sort. The British studies did not assume that therapeutic support for families would help them cope any better than on their own, with what one might think of as native coping strategies and support networks. Given the transatlantic divide in the cultural place of – and values ascribed to – therapy, in general, observed by historians and social anthropologists, the circumspect approach towards offering therapy to families taken by British psycho-oncology is perhaps not surprising.

This difference in the remit of clinical studies between the USA and Great Britain alerts us to a major problem for psycho-oncology: there is no objective definition of 'coping' or of 'a coper' that can serve to calibrate between varying descriptions of, or perspectives on, psychological well-being. Coping strategies vary so much from culture to culture (by which we mean to include nationality, ethnicity, religion, class, political persuasion, degree of urbanisation, and so on, down to the smallest subdivisions of 'cultural identity') that it has proved impossible to reach agreement on what coping is; there is no 'thing' that can be referred to, no shared set of behaviours that people from varying groups share or have shared as examples of coping.

Although psycho-oncology emerged as a subspecialty of psychology in the 1970s, it did not flourish as a discipline in its own right until the 1990s.[4] The field began to expand in the 1980s, developing ways of quantifying adjustment and quality of life among children with cancer and their families. A review of the literature reveals that as the deficit model gave way to the new focus on the strength of families and their coping skills, the topics studied by psychologists diversified, even beginning to acknowledge the possibility of beneficial psychological outcomes. As the subject of study changed from behavioural problems to the quality of life of patients, parents, and siblings, authors began to describe the ways in which the cancer experience could, in some cases, improve people's sense of themselves (Eiser, 2004, p. 28). However, this new psychology of coping has had limited clinical impact on the treatment offered to families of children with cancer thanks to two major and interconnected methodological problems. The first of these is that defining – let alone measuring – quality of life is not straightforward. As Carsten Timmermann has shown, there have been many attempts to develop techniques for quantifying quality of life since the 1970s, but as there has been no single set of values on which patients have agreed, this has been unsuccessful: what counts as a good life depends on age, social connections, and other cultural and material circumstances and expectations (Timmermann, 2012). Without agreement between psychologists on what is being studied, or on how to quantify it, it has proved practically impossible to challenge or build upon studies in the field of psycho-oncology. The second issue is the lack of success in developing predictive models: no reliable linkages have been established between particular behavioural traits and long-term mental and emotional well-being, leaving medical professionals on the front line none the wiser when faced with a child or family struggling to manage the rigours of treatment.[5]

That no accurate predictions have been generated regarding who would flourish and who would suffer long-term psychological problems as a consequence of cancer and treatment is perhaps not surprising: limited predictive power has characterised many subspecialties in clinical psychology. Very few clinical trials of intervention strategies have been run in Great Britain, affording limited opportunities to measure the success of psychological 'preventative' therapy. In the USA, the National Cancer Institute (NCI) has funded many studies of survivors, looking for correlations between personality and health, and treatment and adjustment, but evidence for robust connections has remained scant and inconsistent. Difficulties in agreeing on what is being measured, and on what preventative therapy should cover, echo the difficulties faced by the cooperative groups testing chemotherapy agents in children with acute leukaemia in the late 1950s and early 1960s, prior to the NCI-sponsored conferences at which disease, remission, and response definitions were quantified.[6]

A further problem limiting the ability of psycho-oncological studies of survivors to lead to new interventions has been the longstanding divide between psychologists working with children and those studying adults. The transference of techniques or models from studies of chronically ill adults facing uncertain outcomes to analyses of children's coping has been rare. Instead, paediatric psycho-oncologists have aligned themselves more closely with paediatric oncologists and paediatricians in general, prioritising the patient as a child first and a sick person second. Children living with chronic illness or uncertain prospects have been viewed as facing additional challenges because the progressive emotional and social maturation of children can be disrupted by powerful or prolonged illness experiences. As children with cancer have not been viewed as small adults, techniques for defining and quantifying quality of life have tended to be imported from child psychiatry, rather than from psychological studies of adults with cancer (Chris Eiser, 2004, personal communication).

The quantification methods initially used to assess psychological well-being were those of the American psychiatrist, Thomas Achenbach. His 'Child Behavior Checklist' stemmed from attempts to systematise his recording of paediatric psychiatric symptoms in the late 1960s, and evolved over many decades (Achenbach, 1966). However, having been designed to meet the needs of children he saw in his psychiatric clinic – children with evident mental health problems – the list was extremely negative in tone, focusing on what children could and could not do, and how this might indicate psychiatric dysfunction. The influential child psychiatrist, Michael Rutter, developed a checklist for use in British hospitals to assess children's mental health, but as the theoretical underpinnings lay, once again, in psychiatry, this was similarly rather negative in its assumptions, with questions designed to identify the nature of some underlying pathology.

Through the 1980s, people working with children with many forms of chronic illness cast doubt on the suitability of these checklists for children being treated primarily for bodily illnesses. Open-ended or positively framed inquiries, therefore, designed to elicit information about healthy and exemplary mental and emotional functioning, began to be included. The lists in use since the early 1990s have actively sought evidence of good psychosocial adjustment, a move that has led to even greater variation between the findings of different groups of psychologists and different groups of patients and families.

Local Solutions

Locally available resources and the needs of individual families have profoundly shaped work in Great Britain with families struggling to cope with remission. Paediatric oncologists in charge of some of the largest treatment centres of the 1970s and 1980s have described the provision and evolution of psychological support on their wards.[7] What could be offered to families has depended on local connections between paediatric medical teams and child psychologists and psychiatrists, and the availability of social workers with an interest in and aptitude for counselling. Great Ormond Street and the Manchester children's cancer ward at Pendlebury were once again the pioneers in psychosocial support services for families. It is no accident that the earliest British studies of how families coped were conducted in these two treatment centres and that in both cases American expertise was utilised in the process of building a native structure. However, over the last ten years or so, British teams distributed across the treatment centres have embarked on long-term studies of families from diagnosis onwards, in the hope of finding correlations between early responses and eventual mental and emotional well-being; the optimism that predictive power can be attained remains (interview with Tim Eden, 2004).[8] Work has focused on the length of time anxiety and depression persist in patients and parents, and those involved have begun to address patients directly, rather than attempting to gauge their mental states from parent-completed questionnaires (ibid).[9]

Great Ormond Street's oncology ward was served by a local authority social worker from the start of the 1970s, with an additional social worker being provided by the Malcolm Sargent childhood cancer charity a few years later.[10] While these individuals secured grants and respite care for families, psychological support was provided by clinical psychologist, Richard Lansdowne, and, from the mid-1970s, Christine Eiser. Play therapists were employed and psychosocial ward meetings were held on a weekly basis to discuss those families apparently in need of greater levels of assistance. A peer support group for parents was also founded, again early in the 1970s. By the end of the decade, staff on the ward were conducting studies of how children and their families lived with successful treatment, constructed on the model of a similar survey undertaken at St Jude's Hospital, Memphis in 1975 (interviews with Judith Chessels, 2004, and Jon Pritchard, 2004).

In Manchester, Pamela Barnes founded the first hospital play scheme in North West England in 1971, play therapists having previously been largely confined to the London hospitals.[11] Not only were American ideas imported to Manchester in the 1970s, so were American personnel: social worker, Toni Peroni, was 'borrowed' from Farber's clinic in Boston in order to establish a native tradition of 'total care'.[12] It was clear by the late 1960s, as increasing survival times were being seen across the range of childhood malignancies, that many parents had major psychological problems dealing with their children's fluctuating conditions. Parents and patients seemed reluctant to discuss their worries with oncologists, and when quizzed about this silence, reported not wanting to distract doctors from the business of planning and administering biological treatment (interviews with Pat Morris Jones, 2004, 2005). Morris Jones, by then a Senior Lecturer in the Department of Child Health at the Victoria University of Manchester, approached the head of the department, Professor John Davis (Professor Gaisford's successor), about making use of the skills of psychologists or psychiatrists to help families and identify effective approaches. Peter Maguire from the Department of Psychiatry, whose work with Jean Comaroff we discussed in the previous chapter, was brought in. For the next ten years he collaborated with Morris Jones on research into ways cancer patients and their families could be supported. Maguire left this research area to work with adult patients, taking a Chair in Psychological Medicine in Manchester funded by Cancer Research UK in recognition of his contributions. The studies he undertook were highly influential in establishing medical truth-telling as standard, even within paediatric practice, and in framing psychological reactions to cancer within a broader sociocultural picture, as opposed to purely the product of individual personality factors. These studies stressed the *social* in psychosocial research, an approach that has historically found more favour in Great Britain than in the USA.[13]

Defining 'Cure'

That a cancer diagnosis carried extra weight over and above many other life-threatening or chronic illnesses, owing to cultural perceptions of the disease, had long been recognised by paediatric oncologists (*Cancer* 1975a, 1975b).[14] A major component of this was the difficulty of defining 'cure'. Given the inability of oncologists to detect residual disease, and the length of life children would normally have expected, the term 'cure' was rarely used within paediatric oncology before the 1960s. In 1964, Joseph Burchenal at the Sloan Kettering Institute began a major survey of long-term survivors of leukaemia in the USA and Great Britain. From the results he made the bold suggestion that those who had survived for more than 11 years from diagnosis could be considered cured. Roger Hardisty at Great Ormond Street Hospital conducted a similar survey in 1969 with comparable results. The majority of practitioners, however, were still uncomfortable with the term. A more robust definition emerged in the 1970s: the statistical cure. By studying relapse rates over time, cure came into effect when the risk of relapse was statistically determined to be less than one or two per cent. This entailed a wait of many years before the success or failure of treatment could be established, a situation that for patients and their families, as well as their oncologists, was far from satisfactory. Faced with this uncertainty, patients, parents and psychologists sought to develop alternative measures of success.

While the transformation of culture-wide attitudes has, necessarily, been a long-term target, a possible solution to the uncertainty of remission experienced by individual patients was the concept of psychological cure, first fully described in the mid-1970s by the paediatric oncologist, Jan van Eys. van Eys (1977a) was critical of medical personnel who shied away from the term 'cure', reasoning that the very definition of childhood presupposed a future; therefore, to act as if a child would not survive was tantamount to 'psychological euthanasia'.[15] He argued that clinicians should treat all children as if they would be cured in order that those who did survive could develop normally. Even staff and patients at van Eys' clinic, however, had reservations about redefining leukaemia as measurably curable at the psychological level. At the close of a 1976 workshop held at the Houston paediatric cancer unit of which van Eys was director, there was a determination to avoid the word 'cure' from all parties, including patient and parent representatives, owing to the high physical and psychological costs borne by survivors. Participants wished it noted that 'only Jesus cured' (van Eys, 1977b, p. 129). van Eys was not easily discouraged, however, and urged his audience at a 1980 NCI conference to be less apprehensive of thinking and talking about cure:

> The medical community vigorously pursues biological cure and hedges on conceptual psychological cure. Yet we cannot guarantee biological cure, but we can generate psychological cure if we believe in it ourselves (1981, p. 39).

Other clinicians also frequently spoke and wrote about the need to attend to families' sense of 'curedness'. The textbooks of the 1970s noted increasing cure rates, but chapters on offering psychological support to parents only gave suggestions on helping families cope with terminal illness.[16] Through the 1970s and into the 1980s, to compensate for the lack of advice in manuals and most training programmes, leading clinical researchers in the USA and Great Britain took every opportunity to promote a more optimistic view. In papers and public lectures covering the improved survival rates from recent clinical trials, it was constantly emphasised that families needed to live in anticipation of cure. It was time to accept many childhood cancers – leukaemia in particular – were curable: in a review article of 1985, Donald Pinkel dismissed previous ways of evaluating leukaemia treatment protocols by comparing their 'seven-year cure rates', as belonging to 'the *pre-curative* era' (1985, p. 92, emphasis added). Families were to be encouraged to act, throughout the treatment process and afterwards, as if cure and normality were possible. Interviews with parents and former cancer patients, however, have repeatedly revealed resentment at being told to act 'normally', as if this were possible in the face of demanding treatment protocols and the threat of death and long-term disability.[17]

Through the late 1970s and early 1980s, increasing numbers of medical professionals came to share and espouse the belief that biological cure was possible, yet for many patients psychological cure remained elusive. During the 1980s, van Eys repeatedly addressed the problem of 'how to avoid creating a biologically cured child who remains a psychosocial cripple' (1985, p. 160). What was needed was work towards the 'endpoint of the circle' (van Eys, 1987, p. 118). The child cancer patient needed to be able to 'transcend survival' (ibid); incorporate cancer as a past event that did not threaten his or her being (van Eys, 1985, p. 162). If former patients could not achieve this, they would remain outside society, 'survivors rather than participants' (van Eys, 1987, p. 118). By the late 1980s, some paediatric oncologists and most psycho-oncologists were promoting a finite patient pathway as the ideal, where cancer was to be taken as an episode with a clear end, for survivors as much as for those who succumbed to the disease. This ideal trajectory, however, has not matched up against the cancer experience typically related in patient memoirs. The finite pathway concept has also been in tension with the actuality of treatment plans for many patients, where programmes of life-long follow-up checks have limited survivors' capacities to 'move on' and away from cancer by repeatedly reminding survivors of the importance of their journey through cancer.

Numerous resources were developed to help families get through the period of future-less time. Social support designed to facilitate 'normal family life' and 'better coping' became standard in clinics in Great Britain and the USA through the 1980s. Voluntary bodies, such as The Malcolm Sargent Cancer Fund for Children in Great Britain and Candlelighters in the USA, expanded their work from supplying social workers in treatment centres to offering counselling to individuals and groups of patients and parents. Their efforts were amplified by the foundation of other voluntary bodies that funded play workers in hospitals and holiday accommodation for affected families.[18] Voluntary bodies, hospitals, and oncologists' professional bodies commissioned academic sociologists and psychologists to produce leaflets for parents on how to cope with living in remission. The psychosocial needs of children and their families were met with waves of money for projects designed to relieve stress and increase coping.[19]

Survivorship: When is a Cancer Patient not a Cancer Patient?

The first use of the term 'cancer survivor' is generally attributed to Fitzhugh Mullan, whose 'Seasons of Survival: Reflections of a Physician with Cancer' was published in the *New England Journal of Medicine* in 1985. Mullan divided survival into three distinct 'seasons' of survivorship: acute, extended, and permanent. The following year, together with representatives of many US organisations, Mullan founded the National Coalition for Cancer Survivorship (NCCS), to advocate for survivor's needs and challenge the widespread use of the term 'cancer victim' (Bell and Ristovski-Slijepcevic, 2013, p. 409). The original aim of the NCCS as a body, and of its use of 'survivor', was to empower patients, and surveys in the USA have found the majority of those who have undergone treatment for cancer have embraced the term.

The concept of cancer survivorship never gained the same popularity in Great Britain. A 2012 survey of ex-breast, ex-colorectal, and ex-prostate cancer patients found the majority rejected its use. Some respondents felt their experience did not warrant use of the term, perhaps reflecting the improvement in survival rates since the 1980s, while others objected to the implication that their survival had been a choice: that they had won the battle (Khan et al., 2012, p. 180). This was seen as disrespectful to those who had not survived. Some were reluctant to embrace the term given that their cancer could recur (ibid, p. 181). Many of the respondents did not see themselves – and did not wish to see themselves – as being defined by their cancer experience (ibid, p. 182). This was a survey of older individuals, with a strong sense of their own identity, who viewed cancer as just one of many events in their lives. For childhood cancer patients, still in the process of discovering and creating their own identity, and with far fewer experiences to rate alongside that of the disease, cancer inevitably has had a greater impact. The experience has not only taken up a larger proportion of life, it has also occurred during a period of vital development and socialisation, further problematising the difficult process of 'moving on'. And the routine strong encouragement for former child patients to remain in long-term follow up for the rest of their lives, attending periodic health checks at specialist cancer clinics, also increases the likelihood that the experience of having had cancer will remain part of self-identity for these survivors. But surveys of large numbers of childhood cancer survivors exploring how members of this group feel about defining themselves as 'survivors' have yet to be undertaken, or at least published.

Late-Effects

The extent to which a childhood cancer patient can leave the physical and mental consequences of their experience behind is debatable. The physical effects of treatment, if not permanent, can take many years to remedy. The possibility of 'late-effects' – side-effects that become apparent a long time after treatment has finished, and that may themselves be life threatening – also hangs over families, as do the practical limitations that are the long-term gift of cancer survivorship, such as difficulties securing travel insurance, health insurance, or life cover and mortgages.

The long-term effects of cranial radiation on cognitive function have long been recognised. A recent study on educational attainment by the Centre for Childhood Survivor Studies at the University of Birmingham found the only statistically significant deficits were among survivors of central nervous system tumours, who necessarily undergo cranial radiation, and of leukaemia. Owing to these complications, prophylactic cranial radiation in leukaemia treatment has largely been replaced by chemotherapy administered through the spinal canal since the early 1990s (Lancashire et al., 2010).

The physical effects of treatment can also impact on a patient's sense of identity. A 2007 Australian study found the predominant concerns of male childhood cancer survivors were issues of physical fitness. For female survivors, such consequences of treatment as scarring, weight loss and gain, fertility worries, and damaged hair growth could have an extreme mental cost. A number of the study's female participants suffered from eating disorders, bulimia in particular, apparently attempting to master a body that had so severely let them down (Drew, 2007). A 2009 study by the Late-effects Group based in Sheffield, England, found reproductive issues, memory problems, depression, mood swings, and weight gain to be widespread concerns, alongside more specific physical effects as poor eyesight, hearing problems, and difficulty breathing (Michel et al, 2009).

Recognising the significant trauma that could result from a child being diagnosed with cancer, the American Psychiatric Association broadened the taxonomy of post-traumatic stress disorder for their fourth Diagnostic and Statistical Manual of Mental Disorders (DSM-IV), published in 1994, to accommodate both being a child diagnosed with cancer, and being the parent of such a child (Bruce, 2006). Many studies of childhood survivors have reported decreased self-esteem and confidence, increased anxiety and depression, sleeplessness, and guilt: survivors' guilt that they were still alive when many of those they had met during treatment were not, and guilt for what their loved ones had had to endure (Cantrell and Conte, 2009; Wakefield et al., 2010).

In the spirit of the times, researchers have begun to acknowledge positive consequences of the cancer experience, including a deeper appreciation of life, less antisocial behaviour, stronger bonds with family, and 'psychosocial hardiness' (Wakefield et al., 2010, p. 270). However, as a number of these studies admit, interpreting the positivity or negativity of a consequence is not straightforward. While the behaviour of childhood cancer survivors has been reported by teachers to be less aggressive and disruptive than their classmates, the same teachers have also described them as less popular than their peers, facts that may not be unrelated. Similarly, that survivors have understandably tended to have stronger bonds with their families could also have repercussions. A number of studies have found survivors' rates of marriage and cohabitation to be lower than their peers. It has been speculated this could be a consequence of intense familial bonds impeding the development of their autonomy and independence (Dietz and Mulrooney, 2001).[20.]

A 1998 paper employed the concept of liminality to describe the effects of the cancer experience. Originally used by Arnold van Gennep in his 1909 *Rites de Passage* to describe rituals signifying transitions through the cycle of life in traditional societies, the term was 'rediscovered' by anthropologist Victor Turner in the 1960s. Turner used it to describe the many instances, in all societies, when a person experienced an 'in-between period' – a situation where the individual might be separated from the rest of society, with normal structure and routine suspended. The paper's authors claimed the labelling inherent within the cancer diagnosis itself was enough to induce liminality, and identified three main themes of the experience that intensify this effect: *Cancer Patientness*, an immediate and persistent self-identification as a cancer patient; *Communicative Alienation*, an inability to communicate the experience with social familiars; and *Boundedness*, the way the patient's world contracts, as he or she becomes aware of the limits to space, time, and empowerment (Little et al., 1998, p. 1486). 'In-limbo' status appeared to be confirmed by a 2009 study, in which the authors described the paradoxical situation in which many childhood cancer survivors found themselves. While wishing to be treated normally, childhood cancer survivors also felt a strong need for their experiences – and the ordeal they had gone through – to be verified and validated by others. The same study explained the burden others often placed on survivors, by expecting them to live with a sense of 'perpetual gratitude', celebrating each day, and the guilt induced when the survivors, understandably, failed to live up to this (Cantrell and Conte, 2009, pp. 317–18).

The Patient's Voice

Published cancer narratives began to appear in the 1970s, and have proliferated since then, particularly in the last decade of the twentieth century. These first-person accounts were initially heralded as an important counterpoint to the grand narrative of biomedicine, an authentic, subjective patient voice. Since then, however, analyses of narrative have increasingly questioned these assumptions. The naivety of accepting narratives, particularly illness narratives, as direct conduits to phenomenological truth, and the need to approach and evaluate them within context – particularly with an awareness of the 'blueprints' held within the master narratives of the time – have been strongly emphasised (Rimmon-Kenen, 2002, pp. 11–14).[21] As Arthur Kleinman, one of the earliest to treat illness narratives as a worthwhile subject of study, has stated, the ill are 'somewhat like revisionist historians', refiguring the past in light of the present, often retrospectively imposing a coherent, narrative structure in an attempt to tame the events: transforming a 'wild, disordered *natural* occurrence' into a domesticated and mythologized '*cultural* experience' (Kleinman, 1988, p. 48).

The usefulness of the narrative form for exploring the cancer experience, however, is not in doubt. Although published narratives by survivors of childhood cancer are rare, they have been used by various researchers as windows through which to study the lived experience of coming through the disease. A Canadian study published in 2006 found the overwhelming theme of 39 survivors' narratives to be 'life is never the same'. The world was no longer as safe, secure and certain as before, and never could be again (Woodgate, 2006). The authors described a state of ongoing biographical work:

> [survivors] ... occupy a fundamentally changed personal and social identity that will, to varying degrees, have a lifelong effect on their personal wellbeing and the nature of their participation in social living. Survivors' stories reflect ongoing efforts at cancer-related biographical revision, self-reconstruction and narrative repair, even twenty years after the end of successful treatment (ibid, p. 284).

The 'epidemic' of survival among childhood cancer sufferers witnessed from the end of the twentieth century, as Sarah Drew observed, is uncharted territory, not only for the patients themselves, but for the medical community and society as a whole, something the ex-patients in her study recognised:

> Participants accepted the nature of their position as part of the first generation of survivors following cancer in childhood, and acknowledged that the area of long-term survival is still being investigated and understood (Drew, 2007, p. 287).

The 2009 study by the Sheffield Late-effects Group mentioned earlier found the topics childhood cancer survivors most wanted to discuss with medical practitioners to be their current health, late-effects, medication, and fertility issues (Michel et al., 2009). However, a common theme in worldwide studies has been a sense of abandonment (see Woodgate, 2006; Drew, 2007; and Cantrell and Conte, 2009). The medical community, which has played such an important – perhaps overwhelming – role in their short lives, has been struggling to adapt to the needs of this first generation. Survivors have often been uncertain where to turn to access whatever information and support may exist. Their treatment may be over, but for many childhood cancer survivors, the dialogue and support needs to continue.

Conclusion

Childhood cancer is a transformed and transforming illness. The earliest known survivors of cancers requiring chemotherapy were treated in the 1950s, and few have lived through their natural life spans; our understanding of the long-term consequences of being treated for cancer in childhood using modern techniques is therefore still in its infancy (Laszlo, 1995). Survival rates in Great Britain have leapt dramatically from around 30 per cent 40 years ago to over 70 per cent today, calculated across all cancer types affecting children under the age of 16 (Children's Cancer and Leukaemia Group, 2012). Almost all children currently being treated are enrolled on one or more clinical trials seeking to discover even more effective, or less toxic, treatment regimes, and to reduce the burdens placed on survivors by side-effects affecting the body or mind. The experiences of patients past can in no sense be taken as a guide to the results that future patients can expect. The rapidity and constancy of change in childhood cancer outcomes makes for a particularly interesting historical window into how doctors and their patients have navigated through changing expectations.

Before the Second World War, childhood cancer was a rare and usually hopeless condition. Few doctors could ever expect to see a case in their consulting room or ward rounds, and, with hindsight, it is clear that many children suffering tumours, and most children afflicted by leukaemia, died without ever receiving a diagnosis of cancer. In Great Britain, paediatrically trained radiotherapists, Edith Paterson and Ivor Williams, became interested in the studies emerging from New York that revealed cancer behaved peculiarly in the young. Surveys of British cases confirmed the American results: childhood cancer was a different disease from cancer in the adult population. Specialised treatment plans were tried and tested in the handful of patients referred to the small number of experienced cancer doctors in Britain's biggest cities. Paterson and Williams published the results of their analyses of cancer incidence and of their treatment regimes, sharing expertise across the Atlantic with the small numbers of interested physicians working in cancer centres in America.

In the wake of the Second World War, medical statistics reached a new maturity as counting machines that could manipulate data drawn from large populations became available, and visionary experts studied experimental design to pose powerful questions answerable with the new technology. The power of medical statistics magnified the potential of clinical trials to address queries about treatments' efficacy, and multiplied the scale of epidemiological studies seeking connections between life events and sickness. At the same time, concerns were growing about the health effects of radiation, and the risks posed by the atmospheric testing of nuclear weapons, by locally high levels of environmental radiation, and by the increased use of radiation in medicine. A surge in the number of cases of acute leukaemia being seen in children under the age of five galvanised the Medical Research Council to support nationwide epidemiological investigations into the possible causes of childhood cancer, and small-scale trials of new potential chemical treatments for leukaemia.

American research in the 1960s explored the effects and limits of combining chemotherapeutic agents, first to combat leukaemia and then in the treatment of solid tumours in conjunction with surgery and radiotherapy. The results of the clinical trials conducted by the newly constituted cooperative research groups began to transform the status of childhood cancer from hopeless to curable. Surgeon Denis Burkitt published startling results showing that a highly aggressive and inoperable tumour could be halted in some cases by a handful of pills. Chemotherapy seemed to hold enormous promise for children with cancers of all types. Children with cancer were thrust into the medical spotlight and the public eye, as it appeared that many lives could be saved if only Great Britain could make the resources available.

For the first time, in the 1960s British newspapers carried stories about individual cancer patients and the battles their families faced trying to maintain hope in the face of restricted options and limited resources. Memoirs were published detailing the treatments administered, their side-effects and the alternatives being marketed by practitioners operating outside the mainstream medical networks. Popular scientific journals dedicated many pages to the exciting new fruits of research from cancer centres in the USA, Great Britain, and elsewhere.

By the end of the decade, arguments were being fought in the medical press and in government over whether a system of centralised care and specialist facilities to extend the benefits of research to all children in the country should be funded by the state.

In the 1970s, paediatric oncologists formed first local research groups and then a national professional body to increase access to experimental drugs and the protocols for their administration in sequence and/or combination, and to pool experience of delivering the treatments and knowledge of the outcomes for patients. The number of British children receiving the best known treatments increased from a few per cent towards the current level of over 90 per cent; many of these children, then and now, have also contributed to the ongoing programme of clinical trials to improve results for future cancer sufferers. As more children were being cared for in specialist centres, the psychological needs of them and their families became more obvious. Supportive care came to encompass play therapy and social worker support alongside transfusions of red cells and platelets to keep children fit for life for as long as life was possible.

As Chris Feudtner (2003) showed in his study of the story of diabetes in the middle years of the twentieth century, when an illness is transforming it is vital that the medical profession acknowledges the additional and peculiar burdens this can place on patients and their families. In the absence of certain or even likely outcomes, and when the paths trodden by past patients cannot be held up as examples of likely futures, patients and families ride high and stormy tides of hope and despair with each new finding or anecdote. Sufferers need the support of their medical teams to understand the ways in which their choices and hopes are altered, and continue to alter in the face of increased knowledge about the disease process under new treatments.

The majority of cancer 'cures' are still uncertain; very few cancer physicians or surgeons speak of cure without qualification, as there is always a risk the cancer might return, and a patient can only truly be known to have been cured once she or he has died of unrelated causes. This situation is unlikely to change in the foreseeable future. Medical professionals will continue to present survival data in terms of remission length, five- or ten-year survival percentages, or to otherwise modify the term with adjectives such as 'essentially' or 'in effect'. Such qualifications are particularly unsatisfying when applied to the life chances of a child; where the normal life expectancy might be a further 70 years, a ten-year survival chance is of little comfort.

It is our hope that this historical review of what childhood cancer has been will help shape current debates about what childhood cancer now is and can become. The disease is still seen as an anomaly, an aberration given all we know about the causes of cancer and the nature of childhood. But as the number of survivors continues to increase, and to swell ever faster as a greater proportion of children the world over are able to access curative treatment, childhood cancer may come to seem less strange a proposition. It may become one more chronic health condition that does not provoke public alarm above all other challenges to childhood and does not bring to mind images of pale faces and bald scalps. It may lead to invaluable insights about the modes of operation of other forms of cancer, including those with far higher rates of incidence. Finally, the changed face of childhood cancer may be used as illustration of the enormous transformative effect that even small research groups may have, and as an example of the power of cooperation between medical specialisms, between hospitals, and between countries.

Notes

Introduction

1. The case of cervical cancer and its connection to the human papilloma virus is the best known of these developments.

1 Childhood Cancer: A Disease Apart

1. This story is reconstructed from Tod (1940).
2. The best known of these operations is the radical mastectomy, developed at Johns Hopkins University Hospital in Baltimore, the American beachhead of Germanic surgery; Lerner (2001) has looked at the progressively more ambitious surgeries performed on women with breast cancer during this period; Moscucci (1993) has studied the development of invasive surgical procedures for female reproductive cancers.
3. See Cantor (2008) for chapters detailing discussions of cancer that took place in assorted media in the USA and Great Britain in the first half of the twentieth century.
4. For the development of children's hospitals in Great Britain, as distinct from paediatric departments within teaching hospitals, see Lomax (1996).
5. Sir (George) Frederick Still was Great Britain's first professor of childhood medicine, holding the position at King's College Hospital from 1906, in recognition of his work at Great Ormond Street Hospital.
6. In 1938, the British Association of Radiologists and the Society of Radio-therapists agreed to merge, forming the Faculty of Radiologists, incorporated in 1940. This Faculty was housed by the Royal College of Surgeons. The Faculty's publication, the *British Journal of Radiology*, continued to carry papers on both radiodiagnosis and radiotherapeutics (see Murphy, 1986, pp. 7.4–7.44).
7. Wall (2011) has shown the degree to which St Bartholomew's was unusual in its embrace of laboratory science before the 1920s.
8. Gaisford's influence on cancer services for children in Manchester was profound and could be felt over several decades.
9. See, for example, Williams' presidential address delivered to the British Institute of Radiology in May 1957 (Williams, 1957).
10. One of the most successful centres in the treatment of Wilms' tumours was the Boston Children's Hospital. Early twentieth-century use of surgery alone in this hospital cured less than ten per cent of patients; this was improved through surgical innovations in the 1930s to 40 per cent, and the addition of postoperative radiotherapy in the late 1940s pushed the cure rate up to around 50 per cent (see D'Angio et al., 1980).
11. See Derek et al. (1980) for centre-by-centre figures on the use of radiotherapy alone. The M.D. Anderson Center published its own historical survey of treatment and results (see Fuller et al., 1973).

12. The Royal Victoria Infirmary in Newcastle, the Christie Hospital and Holt Radium Institute in Manchester, the Royal Marsden Hospital in Surrey, and St Bartholomew's Hospital with Great Ormond Street Children's Hospital in London each published research papers in the 1950s or reviews in the early 1960s that provide a wealth of detail about the radiotherapy administered to young patients (see Paterson, 1955; Pearson, 1964; Pearson et al., 1964; Easson, 1966a, 1966b).
13. See Paterson (1955) for a selection of such images.
14. At present, roughly half a million children receive inpatient or daycare surgical treatment in hospital each year, while a further two million receive attention in accident and emergency departments. While it is still the case that the majority of surgeons operate on both adults and children, there are some childhood surgical disorders that are not often seen in adult practice, a situation that has led to the establishment of approximately 130 consultant positions in paediatric surgery.
15. See, for example, the publication of his contribution to a symposium on primary malignant tumours of bone held at a meeting of the Faculty of Radiologists in Leeds in June 1945 (Cade, 1947). This paper presented the results of treatment on 135 cases of bone cancer seen at Westminster Hospital between 1922 and 1944, and the radiation prescribed for each form of cancer.

2 The Rise of Childhood Leukaemia

1. Cooter and Pickstone (2003) provide an overview of the changing levels of support for medicine during the middle of the century; Pickstone (2003) outlines three different types of medicine that were successively and variously developed in the USA, Great Britain, and other countries over the century.
2. For more detail on each of these discoveries see Jolliffe (1993), Feudtner (2003), and Murphy (1986). See also Le Fanu (1999, pp. 238–44) and Bud (2007, Ch. 1).
3. Lewis (1992) and Cherry (1996) both discuss the limitations of the Act, as only the health costs of workers were covered and not those of dependents.
4. Webster (2002, Ch. 1) lays out the difficult negotiations that took place before the NHS could come into being. Berridge (1999, Ch. 2) provides a different perspective on the political changes that came about during the Second World War, which elicited the commitment to a national health service.
5. Thomson (1974, 1976) presents a detailed history of the Council's first 60 years, written by one of its longest standing employees. Austoker and Bryder (1989) provide critical perspectives on aspects of the Council's work in its first 40 years, and Bryder (2011) offers a nuanced account of the Council's activities before 1940.
6. Valier and Timmermann (2008) and Quirke and Gaudillière (2008) explore the mechanisms by which the MRC consolidated state control of clinical research within medicine in Great Britain after the Second World War.
7. Bryder (2011) provides a persuasive account of the move towards statistical and scientific rigour in MRC-funded research in the 1940s.

8. See also Wainwright (1990) for an account of the context of the very first hospital trials of antibiotics.

9. FD 1/245, FD 1/392, and FD 1/393 detail the complex nature of the distribution systems for several of the then novel antibiotics that the MRC wished to have tested by clinicians, while also suggesting the limits of the MRC's ambitions. There were constant worries over the supply and cost of new agents. Paediatric case reports are present in these files.

10. See, for example, Gale et al. (1952): when designing their trial, the investigators had not appreciated the significance of an initial throat swab positive for the bacterial agent.

11. The same doctor, while working as a medical officer for the British Army in the First World War, surveyed the world literature on another form of leukaemia, chronic myeloid, and published on that subject too. Dr Ward (1885–1962) lived in Sevenoaks and was a keen local historian and antiquarian. He wrote a popular local history book called *Sevenoaks Essays* in 1931. Collecting historical data appears to have been a lifelong interest.

12. Dargeon (1940) states that death followed diagnosis between eight days and three months.

13. The Advisory Committee on Cancer Registration had been established in 1969 to examine the state of cancer registration across the country and make recommendations to improve rates of return; the work of this committee included recording the initial aims and subsequent development of the register, a tradition of remembering its own history that the Office has maintained. In light of heightened demand for information about cancer by government bodies and the general public, the Committee was reformed in 1978 in order to review how statistics of cancer incidence were being kept.

14. The paediatric cancer centre in Ohio sent observers to Manchester in 1954 prior to setting up their own register on the same model later that year (interview with Henry Marsden, Manchester, 9 February 2005). Those responsible for the care of children with malignant disease in Newcastle started a register in 1968, again explicitly on the Manchester model (interview with Alan Craft, London, 26 July 2004).

15. The first nuclear device designed to explode as a chain reaction had been detonated on 16 July 1945 in New Mexico, but this was not delivered as a bomb.

16. M. Susan Lindee (1994a, 1994b) gives a good account of the foundation of the ABCC and the politics of the time. Her prime focus is the Commission's genetics project, which had a tighter research focus, looking only for heritable genetic damage from exposure to radiation, than either of the two departments that were concerned with childhood cancer – haematology and developmental (paediatrics).

17. An MRC meeting to discuss the wording of a report on the effects of radiation in patients treated with radiotherapy for ankylosing spondylitis in May 1956 noted that between the years of 1947 and 1954 there had been 94 confirmed cases of leukaemia among bomb survivors, against 25 expected. They also noted that the disease had an average latency period of six years; thus, the total number of excess deaths was expected to rise further (FD 1/8125).

18. This was a paper given in March 1966 at The Conference on The Pediatric Significance of Peacetime Radioactive Fallout, held in San Diego.

19. The file contains the notes of the first meeting of the Committee on Medical Aspects of Nuclear Radiation, held 7 June 1955, where John Freeman Loutit presented a summary of the ABCC's findings. Loutit ran the MRC's radiobiology unit at Harwell.

20. Ryle was also king's physician to George V from 1932 to 1936.

21. Ward (1917) noted that more children under the age of five died from acute leukaemia than did children of other age groups or young adults, but did not draw attention to this and no follow-up studies were conducted.

22. Wailoo argues that by the end of the 1950s, American writing, both in the news and in medical journals, showed the same belief that the increase reflected a rise in the levels of carcinogens in the environment.

23. FD 1/7823. The file contains letters pertaining to the planning for this conference, papers for and minutes of the same, and correspondence from early 1955 relating to the first projects undertaken by the new working party.

24. Leukaemias are currently divided into lymphocytic (or lymphoblastic) and myelogenous (or myeloid) types, the former when malignant changes occur in the marrow cells that would form lymphocytes, particularly B cells, the latter when changes occur in those that would become other white blood cells, red blood cells, and platelets. The acute and chronic types of these two forms make up the four main categories of leukaemia. There are many subcategories and some rarer types that do not fit within the classification system.

25. Minutes of conference, circulated to participants 7 February 1955.

26. The results led to the MRC commissioning a statistically strong prospective study, in which a large group of people would be followed through their lives to see what, if any, effects smoking had on their causes of death. This second, more famous, study entailed writing to all the male doctors in Great Britain in autumn 1951. Two-thirds of respondents – more than 34,000 doctors – replied agreeing to take part in the study, which ran until 2001. Doll and Hill published their preliminary findings, conclusively demonstrating that smoking increased the risk of dying from lung cancer and from heart attack, in 1956.

27. Davis (2007, p. 64) notes that the Hill and Doll studies were not the first to demonstrate the scientific potential of the case-controlled design. The technique was first used by Janet Elizabeth Lane-Clayton in 1923. She compared the life histories of women with and without breast cancer to report on how to prevent the disease to the then British health minister, Neville Chamberlain.

28. It is recorded by Stewart's biographer that she objected to the downgrading of the unit and, by extension, her post (Greene, 1999). Dry (2004) offers a more critical account of Stewart's struggles regarding her professional status. Greene (2011, p. 518) quotes Stewart as saying that the opposition and marginalisation she experienced were 'just right for me personally' and that they made her work harder to demonstrate her position.

29. Letter from Witts to Green, 12 February 1955.

30. The full working party settled upon was Witts as chair, Dacie (a haematologist at the Hammersmith Hospital in London), Court Brown, Doll, Hewitt, Stewart, Green, and Dr M. C. S. Kennedy as the group's secretary.

31. Interview with Gerald Draper and Charles Stiller (Oxford, 19 August 2004) of the Childhood Cancer Research Group, the successor organisation to the Oxford Childhood Cancer Survey, founded in 1975 after Stewart retired. Stewart subsequently extended the survey to all cancers, those survived as well as those that proved fatal, occurring in people under the age of 16. She continued to receive funding from the MRC until she retired owing to the high level of interest in research into the effects of radiation.
32. Minutes of meeting held 19 June 1956.
33. Their data show that incidence rates had not started to decline nine years after the explosions.
34. Richard Doll, 'A Tentative Estimate of the Leukamogenic [sic] Effect of Test Thermonuclear Explosions', presented at a meeting of the MRC's Committee on the Medical Aspects of Nuclear Radiation on 7 June 1955. At the same meeting, nine other papers were presented on occupational, therapeutic, and military exposure to ionising radiation.
35. Letter, Himsworth to Witts, 28 March 1957.
36. Weindling (1992, p. 306) notes that similar distortions in diagnosis rates may well have occurred for tuberculosis and heart disease in children, as these conditions can likewise be 'masked by the onset of rapid and spectacular infections'.
37. See Dry (2004) on Stewart's epidemiological methods and their reception by medical statisticians, and for a highly readable summary of the most significant findings, including those presented by Giles et al. (1956) in the preliminary paper in *The Lancet*. This led to a series of letters from anxious, and in some cases angry, obstetricians and radiologists, through the rest of the year.
38. For an account of the context of Farber's research and its immediate impact on colleagues researching cancer chemotherapy in other centres see Laszlo (1995, Ch. 2).
39. George Weisz (2006) has prepared a major scholarly work on the history of professionalisation and subspecialisation, analysing the trends' roots, spread, and divergence across different nation states. The second half of the twentieth century is covered in more detail in Stefan Timmermans and Marc Berg's study of the development of routinised treatment pathways and professional entry routes in medical subspecialisms (Timmermans and Berg, 2003).

3 Working with Larger Numbers: The Development of Large-scale Clinical Trials

1. See also Quirke and Gaudillière (2008) and Löwy (1996).
2. Cancer chemotherapy research was constructed on the back of the success of a 1945 MRC trial, which demonstrated that antibiotics were more effective than bed rest against tuberculosis. The trial design employed became standard in the USA and Great Britain, its success proving that coordinated effort could reap dividends. Perhaps more importantly, many of the same people were central in the foundation of groups of researchers against first tuberculosis and then cancer, both in the USA and in Great Britain. The link between antibiotics trials and anticancer drugs research is made directly in researchers' memoirs (Taylor, 1990; Laszlo, 1995; Christie and Tansey, 2003).

3. National Cancer Institute (1969) outlines the history of the 32 cooperative groups set up under its auspices, of which 22 were still operating. The review in 1969 marked a move away from these groups being seen as drug-screening bodies towards their being judged solely on the strength of their research. There remained four separate professional bodies to which American paediatric oncologists could subscribe until they combined to form the Children's Oncology Group in 2001: the Children's Cancer Group (originally Acute Leukemia B); the National Wilms' Tumor Study Group (founded 1969); the Intergroup Rhabdomyosarcoma Study Group (1972); and the Pediatric Oncology Group (1979) – the latter an off-shoot of the Southwest Group, formed by paediatricians unhappy with the balance of power between themselves and those researching adult cancers.

4. See, for example, those published in Taylor (1990) and Laszlo (1995), and see Christie and Tansey (2003) for similar recollections from British clinicians involved in the early chemotherapy trials of the 1960s.

5. Frei and Freireich both contributed chapters to Laszlo (1995). Their group and the NCI's chemotherapy and trials programmes have been closely studied by Keating and Cambrosio (2002, 2007, 2012).

6. The rich history of the M. D. Anderson Centre has been covered by James Olson (2009), a professional historian and long-term patient at the Centre.

7. Taylor entered medical education at the age of 34, while working towards a doctorate in psychology at Stanford. Qualifying in 1939, he served as a doctor in the United States Army from 1942 to 1946, chiefly in Korea. He then returned to Duke University as Associate Dean, where he made a name for himself. He was asked to join the ABCC in 1949, as Deputy Medical Director for Research. After three years in this role, he became the Field Director of the Commission (see Taylor papers, series I, box 1, folder 1 (obituaries and curriculum vitae)).

8. Taylor, 'The Beginning of Pediatrics at the Anderson', manuscript in Taylor papers, series I, box 9, folder 141.

9. The sequence of letters between Taylor and Clark regarding the paediatrics department position, running from February 1953 through to July 1953, are in the Taylor papers, series II, box 1, folder 4; letters between Taylor and Mavis P. Kelsey M. D., Acting Dean of the University of Texas Postgraduate School of Medicine, and General James A. Bethea, Assistant Dean, from the same period, are intermixed.

10. Taylor, 'The Beginning of Pediatrics at the Anderson', manuscript in Taylor papers, series I, box 9, folder 141. The manuscript also details how Taylor met with Dr Arthur Kirshbaum in Chicago and discussed moving to Houston and, seeking joint appointments in one another's departments, to facilitate integrated research programmes. Kirshbaum was appointed Professor of Anatomy at Baylor College of Medicine and Professor of Experimental Biology at the Anderson; Taylor was appointed paediatrician at the Anderson and Dean of the Postgraduate School, and a Professor of Paediatrics at Baylor. Kirshbaum, crucially, could provide a mouse colony.

11. Taylor papers, series I, box 10, folder 148, contains a reprint of *Roche Medical Image* June 1968 about the 'Family program for child cancer' at the Anderson, richly illustrated with photographs of the facilities for families in the hospital. Series I, box 9, folder 128, contains a 1973 unpublished manuscript entitled 'Ma' about one parent's involvement with the care of

her child and all others in the same ward, from the time of Taylor's service in Korea, in which Taylor made it clear that this was when he first saw the benefits, nay necessity, of involving parents in the care of their sick children.

12. Taylor papers, series V, box 5, folder 78. The Sutow papers, series III, box 1, folder 11, contains a copy of *The Texan at Houston* from December of 1963 or 1964, which gives a very different account of how six American personnel from the ABCC 'just happened' to end up working together in the Anderson, a story that conceals the large amount of careful planning that went into the coincidence.

13. Taylor's draft of a foreword to a textbook on childhood cancer, dated 7 October 1971, Taylor papers, series I, box 9, folder 127.

14. Dr Edmund Gehan, Head of the Cancer Chemotherapy National Service Center's Biometrics Section, visited the SWCCSG to speak on trial design in the summer of 1962, listing the key questions to address in the planning phase: what do I want to know, how long have I got, how will I measure response, and how generalisable need the result be; Sutow papers, series II, box 1, folder 16; minutes of meeting of SWCCSG held 16 June 1962.

15. William Walter Greulich was the Professor of Anatomy at Stanford University School of Medicine; formerly, he was Professor of Physical Anthropology and Anatomy and Director of the Brush Foundation, Western Reserve University School of Medicine. In the late 1940s, he was working on the manuscript of *Radiographic Atlas of Skeletal Development of the Hand and Wrist* (Greulich and Pyle, 1950).

16. Sutow papers, series III, box 1, folder 2, memorandum dated 3 February 1949. The next folder contains the minutes of the meetings held by this committee, comprised of six heads of department: Mr R. C. Brewer (Vital Statistics), Dr C. A. Harris (Biochemistry), Dr W. J. Wedemeyer (Pathology), Sutow, and Mr Hamako, who advised on the utility of the forms for Japanese speakers.

17. Sutow papers, series III, box 5, folder 60, John W. Wood, *The Present Status of the Pediatric Program*, '1952?' pencilled on cover.

18. Sutow papers, series III, box 3, folder 44, interim report on PE-18 for period 1 July 1953–31 December 1953 shows that during these six months, the clinic saw 1,295 children for the main growth and development study, known as PE-18, and 445 for various other projects and reasons.

19. Sutow papers, series III, box 6, folder 75, contains English translations of many Japanese newspaper articles on the ABCC from the early 1950s. This quotation comes from *Science Asahi* February 1950, a piece entitled 'Visiting the Atomic Bomb Casualty Commission', the section called 'Going Around the Clinic.'

20. Punch card machines were first deployed for analysing census data in the 1890s. The market for, and the types and power of, automatic calculators expanded over the next 50 years. By 1940, punch card machines were available that had been specifically designed for use in scientific projects. See Bashe et al. (1986) for a technical history of IBM's machines and an account of the expansion of the market for them in the 1940s. For the uses to which counting machines were put in Great Britain in the same period see Croarken (1990).

21. The incorporation of statistics into clinical trials during the Second World War was preceded by a quarter century of use in epidemiological studies.

The papers collected in Eileen Magnello and Anne Hardy's (2002) volume recount many of the key characters, theories, and technologies in the development of medical statistics. Magnello's own paper in the collection discusses precisely this claim to greater rationality made by advocates of the new statistics. For the history of the MRC's epidemiological unit from the 1920s to the Second World War see Higgs (2000). On the role of Austin Bradford Hill in shaping clinical biomedical research methods after the Second World War see Rosser Matthews (1995). For the history of epidemiology within the NCI see Parascandola (2001).

22. Sutow was respected by his peers for his development of curative chemotherapy regimens for childhood solid tumours and for his methodological rigour. He was highly cited and frequently asked to write briefing articles on chemotherapy for childhood solid tumours. His input was also requested in many of the high-level policy-making bodies of the NCI. He was, for example, appointed the chair of the paediatric subcommittee of the Solid Tumor Task Force, in 1966. Sutow's status as *the* expert on childhood cancers other than leukaemia in the USA was well established within a decade of his move to Houston. Sutow papers, series II, box 2, folder 45; Sutow was appointed the Chair of this new committee by Zubrod, by letter, 6 May 1966

23. Sutow papers, series III, box 3, folder 32, letter, Sutow to James (Jim) N. Yamazaki, dated 6 December 1954.

24. Taylor papers, series V, box 2, folder 27.

25. Taylor papers, series V, box 2, folder 27, proposal dated 18 January 1955.

26. See for example Sutow papers, series II box 2 folder 45; the Solid Tumor Task Force officially appointed a subcommittee on paediatric neoplasms, with Sutow as its Chair, on 6 May 1966. The aim of this subgroup was to assess data on long-term remissions induced by drugs alone, or by drugs plus radiation or surgery, in childhood neoplasms.

27. Taylor papers, series V, box 4, folder 58, letter to Dr Al R. Shands, Medical Director of the Alfred DuPont Institute in Wilmington Delaware, dated 15 September 1958.

28. Tests conducted on mice in the early 1950s and human patients in the 1960s, demonstrated that using drugs in combination combated the problem of drug-resistant cancer cell development and often produced greater cell kill without added toxicity to the patient. For examples, see Shapiro and Gellhorn (1951), Law (1952), and Frei and Freireich (1965).

29. The group considering typing comprised eight members – Witts (chair), Court Brown, Doll, John Dacie (Hammersmith Hospital), David Galton (Chester Beatty Institute), Ralston Paterson (Christie Hospital), Brian Windeyer (Middlesex Hospital), and Alexander Haddow (Chester Beatty). Dacie and Galton, along with Frank Hayhoe and Witts, were at that stage considered the British experts at diagnosing type of leukaemia, but even they disagreed with one another over a substantial proportion of cases: see, for example, FD 7/323, for an account of the resolution (by compromise) in 13 adult patients admitted to the first clinical trial in 1960. FD 1/7832: the therapeutic trials working party was made up of the same eight men, with the exception of Paterson, replaced by Manchester radiotherapist Eric Easson, and with an additional three members: J. F. Wilkinson (Manchester

Royal Infirmary), E. M. Innes (Edinburgh), and R. B. Thompson (Royal Victoria Infirmary, Newcastle).

30. FD 1/7832: minutes of meeting on 16 December 1957.
31. FD 7/323: minutes of meeting on 9 October 1958.
32. FD 7/314: agenda for meeting 19 March 1959.
33. On the planning in early 1959 for the first trial, see FD 7/314. On the ethics of withholding treatment from the control arm of a trial, see FD 1/7832: minutes of meeting 16 December 1957.
34. FD 7/321: draft of paper submitted to MRC secretariat in July 1962. Bryder (2011) has argued that the early clinical trials conducted by the MRC were designed with the express aim of instilling scientific discipline in clinicians, a position that is certainly borne out in this instance.
35. FD 7/321: letter from Margaret Gorrill to Harold Himsworth dated 4 July 1962.
36. FD 7/314: letter Himsworth to Gorrill dated 12 July 1962. Himsworth was Secretary to the MRC from 1949 to 1968, filling a role akin to that of chief executive.
37. FD 7/323: minutes of meeting of working party for therapeutic trials in leukaemia, 7 January 1960. Children seen by members of the working party were eligible for inclusion in the trial, but, other than Hardisty, members had to contact Doll's secretary, Miss Aylward, in order to have a patient allocated to a specific trial arm. The patients in this first trial came from Great Ormond Street (14 children), Newcastle (11), King's College Hospital (six), St Thomas's Hospital (four), the Hammersmith Hospital (three), Sheffield (one), and St Bartholomew's Hospital (one).
38. See for example FD 1/7830, for Witts' discussions with the MRC staff regarding the plans for a second conference on human leukaemia in the summer of 1957, on the need to explain the idea of randomisation to all participants; letter to Margaret Gorrill, dated 30 January 1959. See also FD 7/314 for Doll's draft patient record sheets, circulated around the therapeutic trials working party on 20 March 1959.
39. FD 7/1337: minutes of conference held 6 June 1957: 'The discussion [of therapeutic trials in leukaemia] was summed up by Dr Green, who said that a pilot trial of a new drug would be quite easy but a statistical trial would be fraught with great difficulty'. The working party set up at this meeting fully understood that its first task was to explore the *possibility* of conducting clinical trials in such a rare and varied disease as leukaemia. The minutes note that 'Agreement could be arrived at on criteria for the evaluation of remission in leukaemia such as the American workers have used at [the NCI in] Bethesda'. These definitions were also to be provided by the working party appointed to explore the possibility of trials.
40. FD 7/323. The second trial opened in 1963 and ran for two years. A third trial ran from 1965 to 1967. A fourth trial did not succeed in recruiting sufficient numbers, and was therefore abandoned when the MRC joined forces with a French group to test immunotherapy in the Concorde trial of 1969. Details of these are in FD 7/325, FD 7/330, and FD 10/218. The trials prior to 1969 have largely been erased from the MRC's internal memory of its leukaemia trials programme, perhaps because there were no cures, or perhaps, in the light of the findings of American groups who were treating leukemic

children 'harder' at the same time because the drugs were given one at a time and in doses now considered too low to be efficacious.

41. See Laszlo (1995) for accounts of the pioneering of these therapies.

42. For comparison, the development of paediatric oncology in the USA has been covered in Krueger (2008).

4   Cancer Microbes, the Tumour Safari, and Chemical Cures

1. Throughout this chapter, we shall use the term for the disease used by those involved at the time – in his writings from 1958, Burkitt named the condition the African lymphoma. However, as cases had been observed outside Africa and not all pathologists were convinced that it was a lymphoma, an international meeting of cancer researchers and clinicians held in 1963 agreed to rename the condition Burkitt's tumour. Further pathology reports indicated that the disease truly was a lymphoma, and from 1970 the name 'Burkitt's lymphoma' began to replace 'Burkitt's tumour'; see, for example, Burkitt and Wright (1970).

2. The figure of 40 per cent comes from The Royal College of Physicians and Oxford Brookes University; MSVA 056: Dr Denis Burkitt in interview with Dr Max Blythe, Oxford, 26 February 1991, Interview Three, p. 14 of transcript.

3. The analysis within this chapter is grounded in the actor network theory of Bruno Latour. This is a sociological theory of scientific knowledge: Latour asserts that what makes a fact a fact is not that it is true, but that it can be made to be accepted as true across a network of places agreed by all interested parties to represent 'everywhere'. Thus, in this account, what led to the present understanding of childhood cancer and of chemotherapy is not that the knowledge claims made in the 1960s by researchers and clinicians were 'more true' than existing and alternative views of these diseases and treatments. Rather, one would demonstrate that the establishment of a new vision of childhood cancer depended, in large part, on the creation of networks of people, places, and ideas, along which the new facts about cancer could travel. In his 1987 book, *Science in Action*, Latour proposed that modern science can be understood as consequent on the same social and political structures as imperialism: science and imperialism both strive to bring things back, unaltered, from a 'periphery' to a 'centre'. In the centre these things (objects, or reports about objects) are recorded, and this process of data collection and combination prompts fresh requests for the periphery to send further data. See Latour and Woolgar (1979) and Latour (1987).

4. Russell bodies are seen in patients suffering from the blood plasma cancer myeloma (Russell, 1890); LeCount (1902) provides detailed instructions for the preparation of samples to reveal Plimmer's bodies.

5. Haviland, Power, Plimmer, and Park all contributed to this special issue of the journal devoted to cancer.

6. It was only with the introduction of powerful magnification equipment, and the development of tissue culture techniques, that 'viruses' were *seen* to exist by influential researchers. Rasmussen (1997) explores how techniques for making the invisible visible changed the nature of biology in the latter

half of the twentieth century. Creager (2001) covers some of the same territory, but focuses rather more on the research fields made possible by new equipment, less on the machinery itself. See also Creager and Gaudillière (2001).

7. Baker offers an insider's reconstruction of the ways in which the relevant NCI committees were structured, and how the NCI's directorate drew in ever greater funding for this branch of cancer research. Gaudillière (1999, 2006) pays particular attention to the work of the NCI on viruses and cancer.

8. The bibliography demonstrates that, until the early 1970s, papers by the leading researchers of cancer germs were published in highly regarded peer-reviewed medical journals.

9. News coverage of Gaston Naessens and his serum for acute leukaemia in 1963 and 1964, covered in depth in Chapter 5, elicited this hopeful claim from some commentators. The simultaneous discovery of virus-like particles in the blood of some patients with leukaemia in 1964 seemed to suggest that while Naessens' approach was doomed, the truth behind childhood leukaemia, and perhaps all cancers, might lie in the virologist's laboratory (Gould, 1964a, 1964b).

10. Anne Marie Moulin (1989) has claimed that the phrase 'immune system' first began to appear in the mid-1960s. Emily Martin (1994) has traced its popular dissemination specifically to a condensed article in *Readers Digest*, published in 1957. See also Tauber (1994).

11. Livingstone (1813–73) worked as a missionary and a medical doctor, spending more than 30 years travelling through Africa, and is credited with revising European understandings of the geography of the interior. He also demonstrated the potential of trading *with* the people of Africa, to replace trading *in* the people of Africa. Burkitt's papers are held at the Wellcome Library in London. Bernard Glemser's (1971) account of Burkitt's work in Uganda reads like an adventure story: the author was better known for his books on popular science for children. There is also an account of Burkitt the 'Fibre Man' by Brian Kellock (1985) but this similarly tells a ripping yarn of medical progress and is of limited value to historians of medicine and scholars of colonialism. Burkitt was interviewed by Dr Max Blythe on four occasions, shortly before his death. The transcripts and videos of these interviews are held at Oxford Brookes University and are accessible by appointment.

12. MSVA 054: Dr Denis Burkitt in interview with Dr Max Blythe, Oxford, 10 December 1990, Interview Two, p. 8 of transcript.

13. No mention of this grant has been located in MRC files. It is likely that it came from the Office of Colonial Development, after recommendation by the staff of the MRC.

14. These files are preserved in the Wellcome Library, and are closed under the terms of the Data Protection Act.

15. This Alexander John Haddow is not to be confused with Professor Sir Alexander Haddow, the internationally renowned chemotherapy expert and director of the Chester Beatty Institute affiliated to the Royal Cancer Hospital in London. Professor Sir Alexander Haddow featured frequently in the British newspapers in the early 1960s, as he was a leading figure in the International Cancer Congress meetings organised by the Union Internationale Contre le Cancer. He was always consulted on the direction

research should take and the significance of results from chemotherapy trials.

16. FD 10/93: letter dated 21 March 1961, A. J. Haddow to Himsworth. Burkitt was at this time in London, meeting Himsworth and Epstein. Burkitt's lecture at the Middlesex Hospital was remembered by Epstein as taking place on 22 March 1961; MSVA 059: Sir Anthony Epstein in conversation with Dr Denis Burkitt, Oxford, 20 March 1991, transcript p. 3.

17. Burkitt used the term 'tumour safari' repeatedly in his correspondence and diaries. He also used it in his published medical writings: see, for example, Burkitt (1962a).

18. The Oxford English Dictionary gives examples of its use dating back to 1860. Hemingway's 1935 novel *Green Hills of Africa* was a bestseller and use of the term 'safari' increased markedly thereafter.

19. WTI/DPB box 1 number 3 (temporary reference), '"Lymphoma Syndrome" Safari in East, Central, and Southern Africa' formal report with financial details, compiled December 1961.

20. FD 10/93: letter dated 25 July 1961, Clifford to Himsworth.

21. FD 10/93: letter dated 16 August 1961, Himsworth to Trafford Smith. This wrangling over the lines of responsibility and credit was not limited to the issue of who should support Burkitt. Similar disputes were recalled in MSVA 059: Sir Anthony Epstein in conversation with Dr Denis Burkitt, Oxford, 20 March 1991, pp. 1–3 of transcript. The full story of the disputes between the three potential British funders remains incomplete because, following the merger of the Cancer Research Campaign (the BECC as was) and the ICRF in 2000, many of the files recording discussions between staff members and minutes of meetings were discarded. The MRC files show only glimpses of the confusion and politicking.

22. London boasted the Chester Beatty Institute and other internationally respected cancer research laboratories, but the strongest claims to 'centre of centres' would be those of the SKI, home to many of the major breakthroughs in chemotherapy research in the 1950s and ever since, and of the NCI in Bethesda, Maryland. Many historians of imperialism have written convincingly that it is a mistake to see power as lodged in the centre in cases where the centre needs what it casts as its periphery to survive and flourish: see Hall (1997, 2000) for an introduction to this exciting body of work. Much British research into drug treatments in the postwar years was conducted in its colonies, and recently historians have asked, what do we fail to see if we assume that the centre controlled these experiments? Most histories of colonialism have paid more attention to the study and management of infectious diseases and epidemics in colonised countries, and less to other medical projects conducted by imperial powers involving their colonised subjects. See Arnold (1988, 1993, 2004), Harrison (1994), Watts (1997) and Worboys (1988a, 1988b, 2004) for an introduction to the history of imperial medicine.

23. British coverage of Burkitt's work was minimal at this date, the first stories appearing in newspapers and general science magazines in 1962.

24. MSVA 056: Dr Denis Burkitt in interview with Dr Max Blythe, Oxford, 26 February 1991, Interview Three, p. 2 of transcript.

25. FD 10/93: research grant application from Burkitt to MRC, stamped received by MRC 20 August 1962.
26. FD 10/93: semiformal report dated May 1963, from Burkitt to Himsworth. Burkitt noted the methotrexate (21 patients) and Endoxan (a trade name for the drug cyclophosphamide; 42 patients) had been given free of charge by the companies developing them. Burkitt (1966) acknowledged the support of Lederle Laboratories, Astra Werke, and Eli Lilly Research Laboratories in a conference presentation of his results trialling methotrexate, cyclophospha-mide, and vincristine.
27. FD 10/93: letter dated 05 December 1962, from Burkitt to Himsworth.
28. Burkitt's diary, dated 17 February 1963, written in New York, mentions a tele-vision film and a piece in a science magazine intended for the general public: WTI/DPB, box 1, number 11 (temporary reference). The same diary mentions a press conference called in his honour, in Dallas, date 20 February 1963. His diary for a longer trip to North America, from September to November 1963, makes many mentions of interviews with the press and a few instances of television interest: WTI/DPB, box 1, number 12 (temporary reference).
29. The diary of September–November 1963 mentions that a British television man visited a private screening of a documentary made by the BECC about Burkitt's research, but not whether or not a public airing was agreed. No cop-ies of the film remain, and there is no record that would indicate that any British television company picked it up.
30. The quote comes from *The Times* (1962), part of a series of articles that month on the work on viruses presented at the Eighth International Cancer Congress held in Moscow. At this meeting Rous was presented with the Gold Medal of the Royal Society of Medicine, and Burkitt's work was applauded. Longer articles that presented the people behind the findings – chiefly Burkitt, the pathologists, and staff from the Entebbe virus research unit – including Burkitt's own articles in *Nature* (1962b) and *New Scientist* (1963).
31. The team comprised Roy Hertz and Min Chiu Li (see Löwy, 1996, pp. 57–8) and Zubrod (1979). Four years later, *The Times* (1960) reported that some of the remissions achieved through 'combination therapy' had been deemed cures, and that this was the first clear evidence of the curative potential of chemicals against cancer in humans. It is estimated that the cancer affects between one in 30,000 and one in 65,000 pregnancies. This type of tumour grows rapidly and quickly leads to metastases in the lungs and brain.
32. *The Times* carried two stories about a new, purpose-built unit opened at the Fulham Hospital in London for the treatment of women suffering from choriocarcinoma. The state-of-the-art ward was kept sterile to protect patients whose immune systems were severely compromised by the treatment. Designed to treat up to eight women at one time, the unit cost £58,000, raised from the Ministry of Health and private donations (*The Times*, 1964b). Six months later it was reported that all ten patients treated had been cured, but the doctor in charge accepted it was too expensive a unit to copy widely, being staffed by 12 nurses, two doctors, seven laboratory assistants and a secretary, all for the benefit of 30–40 patients per year (*The Times*, 1964c).
33. Similar approaches to displaying success against cancer have been observed in other African oncology wards: see Livingstone (2012).

34. The proceedings can be found in Burchenal and Burkitt (1966).
35. MSVA 059: Sir Anthony Epstein in conversation with Dr Denis Burkitt, Oxford, 20 March 1991, pp. 1–3 of transcript.
36. Many theories from the sociology of scientific knowledge stress that scientific experiments typically require the movement of bodies from laboratory to laboratory to make them work, and face-to-face contact before connections between centres of research and agreement can be reached to work together or to work in what can be recognized as being 'the same way'. The case of Burkitt and his viruses, identified by Epstein, exemplifies this. Epstein, in conversation with Burkitt shortly before the latter's death, is recorded as saying: 'you make it sound easy but, if you remember, you actually came to tea with us two days later and we arranged this dispatch of material. But that was not it, because there was an awful lot of trouble after that, and there was a great deal of scientific politics involved, and much above your level and my level, which held it up for what? – I don't know two or three months. And it wasn't until actually I was enabled to come to visit you in Kampala through the generosity of the British Empire Cancer Campaign, as it then was, now of course the Cancer Research Campaign, who funded the trip. It wasn't until that happened that we actually got the thing to work. So even that was fraught with difficulties of a curious kind which led to a lot of trouble'.
37. MSVA 056: Dr Denis Burkitt in interview with Dr Max Blythe, Oxford, 26 February 1991, Interview Three, pp. 12–13.
38. References in the extract are as in the original.
39. MSVA 056: Dr Denis Burkitt in interview with Dr Max Blythe, Oxford, 26 February 1991, Interview Three, p. 5.

5   Making the News and the Need for Hope

1. Krueger (2008) has studied American attitudes to childhood cancer, as derived from surveys of news stories and readers' responses to these and to memoirs written by bereaved parents, from the 1930s onwards. She has focused, in particular, on newspaper coverage of 'glioma' babies – very young children diagnosed with cancer of the eye – and debates over whether or not parents had the right to prevent their children being treated. No comparative study for Great Britain can be made until the 1960s, as we have been unable to locate any news stories or memoirs. A search of articles, obituaries, and advertising about cancer in *The London Times*, for example, produces no articles relating to children, either as individuals or as a patient group prone to distinctive types of malignancy, between 1940 and 1960. It is, of course, hard to prove a negative, but none of the 'broadsheet' press – that is *The London Times*, *The Daily Telegraph*, and *The Manchester Guardian* – published stories before the 1960s. There may have been cases covered in the local press that we have been unable to trace.
2. The foundation of this unit, and its reception by the press, can usefully be compared with the history of funding for, and reactions to, a dedicated unit for childhood cancer research in the Boston Children's Hospital. The initial four-storey Jimmy Fund Building, completed in 1951, was expanded

to eight floors to house research facilities in 1958 and renamed the Jimmy Fund Research Laboratories. From 1948, the Jimmy Fund raised money for research and treatment of childhood cancer, and the charity was well-known in the USA. American charity fundraising has a very different history from that of the British case. For more information on the Jimmy Fund, see Krueger (2008). It would appear that the foundation of the LRF heralded the importation into Great Britain of fundraising tactics and press relations customary in USA since at least the late 1940s.

3. For a history of the largest British cancer care charity, Macmillan Cancer Support, see Rossi (2009).

4. See Chapter 2 for an account of the MRC's 1957 leukaemia conference.

5. Telfer (2008) recounts the history of fundraising at Great Ormond Street Hospital from its earliest weeks of operation.

6. Archives of Leukaemia and Lymphoma Research, the present name of the charity that was founded as the Leukaemia Research Charity.

7. FD 9/1934: letter dated 15 November 1962.

8. FD 9/1934: letter dated 14 November 1962. One of the dates on the initial letter and the draft reply must be an error.

9. FD 9/1934: letter from bereaved mother to Dr Godfrey, dated 13 November 1962. Emphasis in original.

10. FD 9/1934: letter to bereaved mother, dated 22 November 1962. The previous draft used the expression 'no known cure', which was amended to 'no specific cure' before the final draft was sent out.

11. FD 23/1308. Mrs A. Sanderson was handling most of the queries by the summer of 1964.

12. FD 9/1934: letter dated 13 March 1962.

13. FD 9/1934: undated copy of the original letter from the mother to Lord Hailsham; Hailsham's reply states that it was dated 04 March 1963. Emphasis in original.

14. FD 9/1934: letter from Grieve to Morley, dated 18 March 1963.

15. FD 9/1934: letter from Hailsham to the mother, dated 25 March 1963.

16. For instance, Quentin Hogg, the Minister for Science, made the same point in his answer to the House of Commons regarding funding for cancer research: he 'assured the House that as regards the Medical Research Council, research into cancer was in no way hampered by lack of funds'. *Nature* (1964) discusses a question put to Mr Hogg in the House on 17 December 1963. The Minister gave figures for MRC funding of research specifically into cancer of £750,000 for the year 1962–63 and an estimated £950,000 for 1963–64.

17. This was part of Freeth's answer to questions from Dr Stross and Sir L. Ropner on the subject of cancer research on 9 May 1961, reported in a letter to *The Times* by Professor Sir Alexander Haddow (1961).

18. To recap, *The Times* (1961b) reported that Oxford University had been awarded a £25,000 grant from the United States Public Health Service to cover five years of the Survey's operation. This compares with the initial grant of £2,000 from a small charity in 1955 and a follow-up grant of £7,200 from the local hospital trust for an extension of the original survey.

19. The symposium was held at the Cambridge School of Clinical Research and Postgraduate Medical Teaching in August 1964, and its proceedings were published: Hayhoe (1965). As discussed in Chapter 4, the first speaker was

Jean Bernard, brought over from Paris, the second Joseph Burchenal, flown from New York, and the third Burkitt, who visited from his base in Uganda.

20. FD 9/789: memo dated 6 November 1969 from unidentifiable author (initials appear to be S. G. O.) who attended a meeting with Gordon Piller on 3 November 1969 at Piller's request. The author notes: 'Mr Piller told me that the Leukaemia Research Fund had at the moment more money than it was easily able to dispose of'.

21. This newspaper article states that a team funded by the ICRF was sent to the Entebbe Institute in the spring of 1962 to work with the Institute's existing team of virologists. For more on Burkitt's tumour and the links between Burkitt and the East Africa Virus Research Institute in Entebbe, see Chapter 4.

22. *The Times* (1963b) covers publication of the report for work funded 1961 to 1962.

23. The ICRF had just opened its new headquarters, a building in Lincoln's Inn Fields which had cost £2.5 million and was equipped with state-of-the-art laboratories.

24. *The Times* writer also mentioned a new cancer charity, the Marie Curie Memorial Foundation, which had increased its expenditure on research from £1,000 in 1960 to £24,000 in 1963.

25. FD 23/1311: letter dated 31 July 1956. The Lady Tata Memorial Fund was established by Sir Dorabji Tata, an Indian industrialist and philanthropist in April 1932 in memory of his wife, Lady Meherbai, who was diagnosed with leukaemia in 1930 at the age of 50, and died a year later.

26. FD 23/1311. Witts wrote to Himsworth on 30 July 1956: 'I should have preferred to have slept on it a bit longer ... But with Alice Stewart, it is always a case of a red-hot needle and burning thread and it was a fait accompli when I heard of it'. Witts thought the science was sound, but Himsworth wanted a second opinion, as he knew that Richard Doll had some reservations about the work – Doll had not expected to see an increased risk of malignancy across all cancers, only an increased chance of developing leukaemia. Thus, on 1 August 1956, Himsworth wrote to Stewart asking her to send a copy of her draft paper to Sir Bradford Hill, and to Hill to explain why he'd done so the following day. In particular, Himsworth suggested that perhaps Hill could help Stewart be less alarmist and more moderate, to remove statements such as 'Meanwhile, in England, at least one child is dying from leukaemia or cancer each week as a direct consequence of having been X-rayed before birth'. After Stewart's revised draft had been accepted for publication, Himsworth wrote to Hill, saying 'I am glad that you persuaded her to take out what you call the "hooey"': letter dated 20 August 1956.

27. FD 9/789: memo dated 9 November 1972. That the LRF had more money than it knew how to spend was also noted by Julie Neale, in a memo dated 8 January 1970, reporting a conversation about the LRF – its history, current financial state, and Medical Advisory Committee – she had had with Roger Hardisty.

28. FD 9/789: memo by Julie Neale dated 22 January 1970, regarding a meeting between the MRC's Neale and Norton, and Piller held two days earlier, states that the meeting ended with a mutual agreement to exchange information about projects being considered for funding on a fairly formal basis. Neale

wrote 'it was apparent he [Piller] held fairly intense views on what should be done' in leukaemia research. Pilller again approached the Council, seeking a more formal connection between the LRF's Medical Advisory Panel and the MRC; letter to Dr Gray, dated 20 March 1973. Dr Galton, a leading clinician funded by the MRC, made enquiries of those he knew on the Panel, and discovered that Piller was acting alone; Neale noted in a memo dated 7 May 1973, that Galton told her Piller had been reprimanded for approaching the Council in such a way. Neale met with Piller on 18 June 1973 (recorded in memo written two days later), and noted that Piller 'displayed somewhat less of his emotional intensity on this occasion than previously; it was accordingly easier to discuss with him sensibly the question of <u>informal</u> collaboration with the LRF, which he was anxious to make clear was what he had always intended' (original emphasis).

29. These articles were found in a file of letters from members of the public and newspaper clippings: FD 23/1308. His article may also have been carried by other local newspapers.
30. Newspaper articles as before, March 1964, FD 23/1308.
31. FD 9/1934. The press release was held on 22 March 1974. Piller sent a copy of the release to Dr Norton of the MRC three days later.
32. FD 9/1934: memo dated 27 March 1974.
33. FD 9/1934: letter to Piller dated 24 April 1974: 'The press interest in your announcement was in the event very great and, as I think I told you on the telephone, it involved my being got hold of at the weekend and being asked to comment when I had no prior knowledge that anything had been said. It was a bit of a fast ball to deal with, as it were "cold"'.
34. For example, in 1959 *The Lancet* carried an article by the Medical Officer of Health for Caernarvonshire. Reporting on increases in the incidence of leukaemia in those British counties with the highest levels of rainfall, between 1950 and 1957, he suggested that strontium 90 was responsible (Alun Phillips, 1959); see also *Guardian* (1959). Lord Hailsham's secretary wrote to the MRC to ask if they would arrange for a retort to be written, not to come officially from them but to be planted with 'a reputable member of the public': undated memo from the last three days of October 1959, FD 23/1317.
35. The files in question are MH 160/74 and MH 160/75, files of the Ministry of Health. A further small file pertinent to the case is PREM 11/4950, which contains the correspondence exchanged between the Minister for Health and the Prime Minister's office on the matter.
36. The story has been pieced together from newspaper clippings preserved in files MH 160/74 and MH 160/75, opened in December 1963. They come from the broadsheets, the tabloid *Daily Mail*, and the local paper, *The West Lancashire Evening Gazette*. The local paper carried stories of how the boy's sisters and father were managing without him and Mrs Burke. *The Daily Telegraph* (1963) carried a picture of Edward being strapped into a helicopter to fly to hospital for his first treatment. MH 160/75: letter dated 1 January 1964, from J. P. Lewis, Group Secretary to the Blackpool and Fylde Hospital Management Committee, to the Principal Regional Officer for the Ministry of Health, Miss A. E. Earlam, stated that Dr Santonacci was one of the most distinguished physicians in Corsica, and was willing to defy the French authorities by administering Anablast.

37. MH 160/75: letter Lewis to Earlam, dated 1 January 1964, as before. The letter closed 'I am quite happy to leave the matter in your hands' – the Blackpool hospital committee declared itself out of the affair, before any flood of requests could begin.
38. MH 160/74. Donald Brown's arrival was reported in *The Guardian* on 03.01.1964.
39. MH 160/74: Orkney case reported in memo regarding telephone call from the area's MP, Mr Grimond, dated 10 January 1964; letter from E. J. Bowers, clerk to Stevenage Urban District Council, dated 8 January 1964; letter from Elizabeth (known as Bessie) Braddock, Labour MP for Liverpool Exchange (from 1945 to 1969), to Minister, dated 19 January 1964.
40. MH 160/74. The boy, the Duke's third child, was born in 1961 and died in 1964. The illness was not named in the letter, but was probably aplastic anaemia. Barber replied that the serum was almost certainly worthless and it would be better for the family to stick with the treatment they were already receiving: letter dated 28 January 1964.
41. MH 160/74: letter dated 24 January 1964.
42. In the 1964 election, Barber lost his seat, returning to government the following year; Jenkins was elected as an MP. He later became an important member of Margaret Thatcher's government.
43. MH 160/74. His offer was covered in *The Guardian*, *The Times*, and *The Daily Telegraph*. *The Times* (1964e) covered the Institut's offer to test the serum free of charge. Naessens had, several years before, refused to submit his serum to government tests, on grounds of the expense, as he would have had to bear the costs himself.
44. MH 160/74: letter dated 13 January 1964.
45. MH 160/74: report to Chief Medical Officer of the meeting, held 18 and 19 January, and its findings. The costs were borne by Mr D. B. Davie from Stonehaven, who also offered Naessens a salary of £3,000 to relocate to and make serum for use in Britain (*The Times*, 1964f).
46. MH 160/75: undated newspaper articles reporting Mr. Davie's offer and Naessens response.
47. MH 160/74: undated draft report, sent to Sir George Godber of the Ministry by Professor Aujaleu of the French Ministry of Public Health and Population, letter dated 28 January 1964. The letter notes that 'the National Order of Doctors, which is the legal authority responsible for maintaining the rules of ethics and integrity in the medical profession has informed doctors that they cannot make use of the Naessens serum, the complete ineffectiveness of which has been demonstrated, without incurring penalties. There is, of course, much public agitation on the matter, especially in Corsica, where feelings are always more intense than in the rest of France.'
48. MH 160/74: memo dated 23 December 1963.
49. MH 160/74: letter dated 27 January 1964.
50. MH 160/74: news clipping dated 28 January 1964.
51. MH 160/74: telegram dated 29 January 1964.
52. MH 160/74: letter dated 27 January 1964.
53. MH 160/74: memo dated 28 January 1964.
54. MH 160/75: letter from Jacobs to Minister dated 1 February 1964; telegram dated 7 February 1964. The replies to each stated that the British authorities had no grounds to intervene in French due process.

55. MH 160/75: a woman from the outskirts of London wrote to the Queen, on 3 February 1964, plaintively explaining that 'If as I fear he is imprisoned 1000's of children will be refused this hope and although the BMA [British Medical Association] condemn this hope is it worse than their offer which is not to think of tomorrow'. MH 160/75 contains a series of letters from Naessens' lawyer, Dudley Clarke, to Edward Heath, and memos about these and draft replies from the Ministry, beginning 25 March 1964, and ending 12 May 1964.

56. MH 160/75: letter to a Northampton MP, for example, dated 10 March 1964, stated that 'research into leukaemia goes on untiringly ... and although no cure has yet been found modern methods of treatment can in some cases give long periods of remission'. The main British sites for research were identified as Great Ormond Street, the Chester Beatty Institute, and the Christie Hospital.

57. MH 160/75: memo dated 20 April 1964.

58. MH 160/75: text from article in file, dated 9 April 1964.

59. MH 160/75. *The Daily Telegraph* (1965) reported that Naessens was fined the maximum sum of £1,300 for illegally practising medicine and pharmacology. The judge described Naessens as 'a typical half-crazed amateur', and expressed frustration that he could not impose a harsher sentence. See also *The Times* (1965a, 1965b, 1965c).

60. For a glowing account of Naessens' work and theories, see Bird (1991). This is not a scholarly work, but does detail later occasions when Naessens was taken to court for selling what doctors described as quack remedies. Orthodox physicians have no patience with Naessens' theory of miniscule 'somatids' as it is in direct contradiction to the germ theory of disease. No established scientist has ever succeeded in replicating Naessens' observations.

61. Aronowitz (2007) and Lerner (2001) discuss the history of alternative conceptions of and treatments for breast cancer, and how these have been strengthened by developments in orthodox treatments.

62. See, for example, Krueger (2008) on American cases of state-enforced treatment from the 1930s.

63. It retains popularity, partly through its inclusion in Gerson therapy, a regime of dietary changes, supplements, and detoxification treatments designed to correct enzyme deficiencies that are supposed to be the causes of cancer.

64. The United States Food and Drugs Administration estimated in 1975 that around $2 billion a year was being spent on what they derided as worthless tests and treatments (*CA*, 1975).

65. FD 23/1308. The GP was Robert Kernohan of Radcliffe.

6   A New Breed of Doctor

1. National Cancer Institute (1969) summarises the range of cooperative groups that was conducting trials into solid tumours in adults and children.

2. See Chapter 1. Pickstone (2007) covers the peculiarities of British cancer research in more detail and expands on variations in the development of the field within Europe.

3. Sutow wrote to a former colleague, 'Our cancer cases are slowly increasing. We have a good selection of cases coming in. For example, our present

census includes two leukemias, one osteogenic sarcoma, one teratocarcinoma, one soft tissue tumour of the mucous membrane of the mouth (?papilloma), hepatic cell carcinoma of the liver, hemangioma of the orbit and thyroid carcinoma. That plus our outpatient cases give us a pretty wide variety of cases': in Sutow papers, series III, box 3, folder 32, letter to James N. Yamazaki dated 6 December 1954.

4. Sutow was the leading spokesperson for chemotherapy to treat solid tumours in children, arguing it had fewer long-term side-effects than radiation and could deliver good palliation for incurables. His first major publication appraising the potential of 'so-called drug adjuvant therapy' (p. 1588) against childhood tumours was published in the foremost oncology journal, *Cancer*, in 1965, and described his first decade's experience at the M. D. Anderson Hospital.

5. This was stated in several of the interviews conducted for this project with some of the first paediatric oncologists in Great Britain.

6. Scott's other senior registrar in the mid-1960s, Derek Crowther, went on to specialise in medical oncology for adults. For more on Crowther's subsequent career, see Christie and Tansey (2003). Crowther was the first head of the Cancer Research Campaign's Department of Medical Oncology, set up at the Paterson Institute in Manchester.

7. At Boston Children's Hospital, early twentieth century use of surgery alone cured less than ten per cent of patients; this was improved through surgical innovations in the 1930s to 40 per cent, and the addition of postoperative radiotherapy in the late 1940s pushed the cure rate up to around 50 per cent; the addition of chemotherapy from 1958 (actinomycin-D, and then vincristine from 1962) delivered a rate of 81 per cent. On Hodgkin's disease and the transformation in expected outcomes, see DeVita et al. (1978).

8. See The Children's Solid Tumour Group (1981) for references to the series of papers from the late 1960s that demonstrated the efficacy of treating non-Hodgkin's lymphoma in a similar way to acute lymphoblastic leukaemia.

9. It is clear from Smith et al. (1977) that the Royal Marsden had been treating children with Hodgkin's disease since 1941, but no date for the introduction of chemotherapy was given. The paper states that in the mid-1960s, ten children were given the MOPP protocol (mustine, vincristine, procarbazine and prednisone) as pioneered in the USA, and that one child had been given triphenyl-methyl-ethylene in 1941 'with no success' (p. 122).

10. Both interviewees related the situation concerning children with tumours of the bone and brain in the early 1970s. Elsewhere, bone tumours were usually viewed as the territory of orthopaedic surgeons and dealt with in local hospitals, often with, Judith Chessels remembers, 'semi-disastrous results'.

11. Bond (1972) gives 'tumour research fellow' as her job title. Bond (1975) gives an alternative job title, that of lecturer in paediatric oncology – a significant switch, suggesting that the subspecialty was becoming established in London by the mid-1970s.

12. Pickstone (2007) covers the development of chemotherapy, and radiotherapists' reactions to the emergence of the field of oncology, in detail.

13. Barnes (1999) and Barnes (2010) detail the history of health services for children in Manchester throughout the twentieth century, and discuss the role of Pendlebury as a site and of Gaisford as a leader.

14. The Royal Liverpool Children's Hospital closed in 1985 and its services were relocated to Alder Hey. It had opened the year after Manchester's children's hospital and had been the third institution to support a professorial appointment in paediatrics, the post being created in 1930 for Norman Capon.

15. Manchester's children's hospital was founded in 1855, just three years after Great Ormond Street; it was only the third such hospital in the country. Birmingham's children's hospital was one of 11 established in the 1860s (Lomax, 1996, p. 26.) Parsons was appointed to the Chair of Child Health in Birmingham in 1930: see Chapter 1 for further details.

16. Malpas and Freeman (1974, p. 712) used this term with no further explanation (p. 712); it was therefore presumably in common parlance within oncology by this date.

17. The figure of 3.5 per cent comes from a survey of 1,245 patients with acute lymphoblastic leukaemia treated between 1963 and 1967, reported in Medical Research Council (1971).

18. See, for example, the study of long-term survivors conducted for the NCI by Burchenal and Murphy (1965).

19. For an account of the establishment of St Jude's hospital, see Wailoo (2001).

20. This paper offers a summary of the trial series.

21. ...'there are indications that attempts at cure are achieving limited success: deliberate cures in acute leukaemia, though still uncommon, are much more frequent than the chance cures of former years' (Spiers, 1972, p. 473).

22. Campaigns now tend to focus on the rights of children in developing countries to have access to curative treatment for cancer.

23. The article calculated that as the 3,759 children who died from solid tumours between 1953 and 1962 were treated in 601 hospitals, each hospital could be expected to be treating only 0.6 cases per annum.

24. *The Times* (1961b) reported that Oxford University had been awarded a £25,000 grant from the United States Public Health Service to cover five years of the Survey's operation. This compares with the initial grant of £2,000 from The Lady Tata Memorial Fund, a small Oxford Charity, in 1955 – see Dry (2004) – and a follow-up grant of £7,200 from the Nuffield Provincial Hospital Trust for an extension of the original survey, as reported in *The Times*. Stewart's group's preliminary findings were published in *The Lancet* on 1 September 1956 (Giles et al., 1956).

25. It is clear from their use of the term 'paediatric oncology' that this referred to a branch of clinical knowledge not a subspecialty. The passage from branch of knowledge to area of specialisation to career structure has been catalogued repeatedly in modern hospital-based medicine.

26. See Medical Research Council (1978) for a report on the trial, and Lennox et al. (1979) for commentary on its reach and effectiveness at extending survival. Lennox et al. stated that the MRC trial recruited 98 children between 1970 and 1973 out of 313 diagnosed with 'nephroblastoma' (or Wilms' tumour), and that this represented 57 per cent of those eligible. The authors also noted that fully 85 per cent of the 313 received at least four days of chemotherapy, indicating that the value of this form of treatment had been widely accepted by 1970.

27. Lauder and Aherne (1972) show that the Royal Victoria Infirmary in Newcastle began treating children with abdominal tumours in 1957. Bristol

was administering chemotherapy to children with solid tumours by 1971 (Ablett, 2002a, 2002b).

28. Mott et al. (1997, p 1448) contains a graph showing the annual numbers of cases referred to UKCCSG members between 1977 and 1999.

29. The terms 'maintenance therapy' and 'continuation therapy' are used to denote treatment administered after remission has been achieved. Donald Pinkel objected to the term 'maintenance therapy', arguing it was not maintaining anything: without it, children would relapse (interview with Jim Malpas, 2004).

30. See, for example: 'an objective definition of cure has not been possible. This lack is partly because the state of complete remission, based on our current ability to detect residual disease, is not distinguishable from a true disease-free state, and partly because the origin of the disease is not understood ... [but] an operational definition of cure [is possible by studying relapse rates over time]' in George et al. (1979, p. 272).

31. This editorial noted that none of the 293 children treated in Manchester for acute leukaemia between 1953 and 1963 had survived.

32. UKCCSG minutes of second meeting, held 24 March 1977, record 21 members present and six apologies, up from 14 members at the first meeting on 14 January 1977.

33. UKCCSG minutes of first meeting on 14 January 1977 record the decision to approach Oxford Survey about collaboration on a paediatric cancer registry. Registration of cases began on 1 July 1977.

34. Of 16 patients registered, nine had been enrolled in the trial: UKCCSG minutes of meeting 9 December 1977.

35. UKCCSG minutes of meeting 14 April 1978.

36. UKCCSG minutes of meetings 14 April 1978 and 30 June 1978.

37. UKCCSG minutes of meeting 28 September 1978.

38. A meeting was set up with the MRC for 20 January 1978 (UKCCSG minutes of meeting 9 September 1977); a year after this meeting, the MRC had stopped recruiting into trials of therapy for Wilms' tumours (UKCCSG minutes of meeting 12 January 1979); the UKCCSG's own Wilms' trial opened on 1 January 1980 (UKCCSG minutes of meeting 28 September 1979). The MRC had also been conducting a trial into treatment for children and adults with osteosarcoma, which also ended in 1980; thereafter UKCCSG members enrolled patients into a SIOP study (UKCCSG minutes of meeting 21 November 1980).

39. UKCCSG minutes of meeting 28 September 1979.

40. UKCCSG minutes of meeting 13 June 1980. At the meeting of 30 January 1981, a UKCCSG member explained how to obtain methotrexate for the trial at £17.50 per gram instead of the usual £100.

41. Lederle sponsored a symposium in 1978 (UKCCSG minutes of meeting 9 September 1977); Eli Lilley paid for the printing of an information booklet for parents (UKCCSG minutes of meeting 8 June 1979).

42. Eli Lilley and Lederle were giving £500 every 6 months, and Wellcome was offering £1000 for a study into Acyclovir for patients with chicken pox; however, office annual expenses were running at £2,000 per annum, covering the salary of a secretary, and telephone and postage costs, so the group agreed to approach Cancer Research Campaign for a more secure source of

funding for salaries for a full-time research associate and full-time clerical officer and their expenses (UKCCSG minutes of meeting 13 June 1980) This bid was successful (UKCCSG minutes of meeting 29 May 1991).

43. UKCCSG minutes of meeting 1 July 1983.
44. UKCCSG minutes of meeting 7 October 1983.
45. UKCCSG minutes of meeting 1 July 1983.
46. UKCCSG minutes of meeting 26 June 1986.
47. UKCCSG minutes of meeting 27 June 1987.
48. UKCCSG minutes of meeting 15 January 1988 note that a recent edition of the series *World in Action* had focused on the report.
49. UKCCSG minutes of meeting 20 June 1990 record the request from the Royal College of Physicians for work to identify the number of training posts required and where these should be located geographically.
50. UKCCSG minutes of meeting 15 January 1988.
51. The concern over services in the Oxford region was noted for the first time in the minutes of UKCCSG meeting 31 May 1985. Discussions were had with the physicians responsible for the region, reported at UKCCSG meeting 9 September 1986, and the Oxford team were invited to join UKCCSG meetings, minuted at meeting 15 January 1988. UKCCSG minutes of meeting 6 January 1989 note that this invitation had not been taken up, and that members feared this might 'promote isolationism'. At the next meeting, it was noted that Oxford had agreed to appoint a Consultant in Paediatric Oncology (minutes of meeting 22 June 1989). However, it was noted that Oxford did not consider that they would have a specialist unit ready to receive patients sooner than 1992/93 (minutes of meeting 11 January 1991).
52. The Chair of the organisation at this point was Tim Eden (UKCCSG minutes of meeting 17 January 1992).
53. UKCCSG minutes of meeting 17 January 1992.
54. UKCCSG minutes of meeting 22 January 1993.
55. UKCCSG minutes of meeting 26 June 1992.
56. UKCCSG minutes of meeting 17 July 1993.
57. UKCCSG minutes of meeting 21 January 1994.
58. UKCCSG minutes of meeting 21 January 1994.
59. *Out of the Wood* was ready for circulation in 1988 (UKCCSG minutes of meeting 23 June 1988).
60. The bodies formally merged in August 2006.

7   Living with Uncertainty: Three Patients on Trial

1. Kim's mother on being informed his specialist had refused to make any more appointments as nothing more could be done (Smith, 1964, p. 103).
2. At this time, an undifferentiated 'cancer' became a Wilms' tumour, distinguished from other forms of malignancy that might grow in the kidney.
3. Tan had grown up and trained in China, leaving to settle in the USA in the late 1940s and joining the paediatric oncology department in SKI in 1952, where she remained until retirement in 1996.
4. We can deduce that the cancer specialist in Paris overseeing Kim's treatment was Odile Schweisguth who, having trained in Boston with Sydney

Farber, had imported his concept of chemotherapeutic treatment for Wilms' tumour to France in the very early 1960s. As discussed in Chapter 6, she went on to found the earliest professional society for 'paediatric oncologists'. Schweisguth stayed in contact with Tan and the other experimentalists at SKI, ensuring that practitioners in Europe were aware of cutting-edge ideas on the role of chemotherapy in the management of metastatic solid cancers.

5. Morton Whitby, previously a Harley Street urologist, had founded the Cancer Prevention Detection Centre in 1960, having become convinced of the ability of an electrical apparatus to detect both cancer and cancer susceptibility. In 1963 he was found guilty of 'infamous conduct in a professional respect' by the General Medical Council and struck from the register. At the time, the Centre was receiving 10–12 volunteers a day. Morton Whitby was restored to the register in 1966 (*BMJ*, 1963, 1966).

6. All six listed authors were employed at St Bartholomew's Hospital.

7. The three who died lived for 15, 16, and 26 months after diagnosis, respectively – in other words, the least responsive patients in the trial survived almost as long as the most responsive amongst the historical series.

8. The authors identified their opponents as Lederman and Jones (1974).

9. The authors noted that some patients, having developed strong skin reactions after the first three or four weeks, had their treatment suspended for one or two weeks, but did not specify which children were affected (ibid p. 248).

10. Figures for the number of delayed courses of chemotherapy and radiotherapy were supplied and broken down by reason for delay; these included infection, anaemia, and skin rashes (ibid p. 249).

11. See the first chapter of Krueger (2008) for a history of eye cancers and parents' reactions to the loss of an eye by their child.

12. For example, see the surveys carried out by the Christie Hospital (Paterson and Aitken-Swann, 1954; Aitken-Swan and Easson, 1959; McIntosh, 1976). A report of a 1959 lecture on the subject by Mr A. Dickson-Wright, treasurer of the Imperial Cancer Research Fund, appeared in *The Times*, provoking strong opinions in ensuing correspondence. See *The Times* (1959a, 1959b), Adams (1959), Curtis (1959), Francis (1959), Howard H. I. (1959), Howard S. (1959), Jenkins (1959), Nicolas (1959), Pantin (1959), Rouse (1959) and Sinclair (1959).

13. Jane also mentioned the rain storm of 1986, which drenched the South Lakes area with radioactivity from the Chernobyl nuclear disaster, devastating many upland sheep farmers, whose flocks were found to contain levels of Caesium-137, rendering them unfit for human consumption. The radioactivity was found to be particularly concentrated in the silt at the bottom of tarns, places Jimmy often played (ibid, p. 25).

14. For a thorough account of the Windscale Inquiry and its historical significance, see Wynne (1982).

15. Regarding the West Cumbrian excess, Martin Gardner has described the methodological temptations to which some analysts fall prey (or might be suspected of having fallen prey to) when faced with a statistical excess: 'By variations of definition of boundaries it is possible to manipulate, if desired, the numerical size of any reported "excess" … The problem is succinctly

described as "moving the goalposts" or as the "Texas sharp-shooter effect"' (1993, pp. 149–50).

16. Personal correspondence with Janine Allis-Smith (2006).

17. Keith Boddy of the Regional Medical Physics Department in Newcastle General Hospital objected that the programme gave no description of Day's methodology, and that the measurements seemed to be 'highly unconventional and unreliable with considerable uncertainty in relating them to "background"' (Keith Boddy, 'A perspective on the Yorkshire Television Programme: 'Windscale: The Nuclear Laundry'' November 1983, p. 1, in Cumbrians Opposed to a Radioactive Environment (CORE) archive). The claims made by Radford, a former chairman of the Committee on the Biological Effects of Ionizing Radiation of the American National Academy of Science, were disputed by Professor Fremlin, Cumbria County Council's nuclear advisor, who argued that the risk of developing cancer from exposure to radionuclides in house dust was lower than that from smoking a single cigarette. Radford responded with his own calculations, demonstrating that the increase in the risk of developing cancer as a result of exposure to the additional radiation in the Seascale area was instead equivalent to that produced by smoking 20 cigarettes a day ('Radford answers Fremlin on cancer risks,' unnamed Cumbrian newspaper, 2 January 1984, p. 15, CORE archives). K. S. B. Rose, a radiobiologist working for the Atomic Energy Authority in their laboratories at Harwell, prepared a memorandum for governmental agencies facing questions from the public and from journalists about the reported excess, questioning John Urquhart's status. He noted that the YTV team 'called him a statistician. He is a librarian!' (K. S. B. Rose, memo dated 16 November 1983, CORE archives). Urquhart responded pointing out that his contribution had been authorised by the chief medical statistician for Christie Hospital and an independent statistical assessor for BNFL (John Urquhart, letter in unnamed newspaper, 1 December 1983, in file 'December 1983,' CORE archives).

18. The team assembled comprised epidemiologists Dr A. M. Adelstein and Professor Geoffrey Rose (both from the prestigious department located at the London School of Hygiene and Tropical Medicine) and Dr Martin Gardner (working at MRC Environmental Epidemiology Unit), cancer specialists Professor J. S. Orr and Professor R. J. Berry (both from the Middlesex Hospital), and paediatrics researcher Professor M. Bobrow (Guy's Hospital Medical School). Observers from the Scottish Home and Health Department and the Welsh Office were also invited to take part but not contribute to the debate, and, finally, a secretariat of three staff was appointed, drawn from the Department of Health and Social Security and headed by Dr Eileen D. Rubery.

19. On the history of government attempts to reassure the public about nuclear safety, see Wynne (1982). Nicholas Leonard (1984), the London editor of *The Irish Independent*, wrote: 'A new weasel word entered the English language yesterday afternoon: "black-wash". It describes a classic Whitehall whitewash carried out under the chairmanship of Sir Douglas Black'.

20. Parts of the agenda of a meeting held by the Ministry of Health on 22 October 1984 were leaked to the Cumbrian press in the months following the meeting. *The Whitehaven News* (1985) broke the story that the panel had

contemplated testing aborted fetuses and dead babies from local hospitals for plutonium traces to see how the element crossed the placenta, and that parents would not be informed of these tests. James Cutler, producer of the original YTV documentary, went into greater detail about the proposed experiments that would be undertaken on tissue samples collected from dead children, babies, and aborted fetuses in a *New Statesman* article (1985). Cutler quoted Keith Boddy as having said there was no need to inform parents if high levels of plutonium were found in a placenta or dead baby, as 'one might be worrying the parents needlessly'. Other proposed experiments were dismissed as ethically unworkable, namely research into the mechanisms by which radioactivity passed from food to child through the gut. Cutler asserted that one idea had been 'to feed "volunteer"' children with local contaminated shellfish and monitor them to see how much plutonium they had absorbed. This particular research project was rejected as (quoting from the minutes of the meeting) 'there might be ethical problems'.

21. Jakeman stated that 20 kilograms of uranium had been released, not the 200 grams recorded in the official figures given to the Black committee (Spackman and Connett, 1986).

22. The Report stated: 'there is no known human leukaemia virus that could be postulated as contributing to the observed excess leukaemia incidence in young people near Sellafield ... The available evidence would suggest that any virus that played a part in leukaemia induction would do so in a multifactorial manner ... the tumour viruses so far described generally being widespread in any population and only causing malignancy as a late and rare consequence of infection, following some other additional environmental agent' (Black, 1984, p. 58).

23. Letter from Richard Doll to Janine Allis-Smith, 15 May 1989, CORE archives.

24. The minutes of Working Party on Leukaemia Research third meeting, held 26 February 1957 and circulated 16 March 1957, include a report from Stewart on her Leukaemia Survey in Childhood. The minute reads: 'There was a general discussion on Dr Stewart's report. Dr Doll said that the main result which seemed to be emerging from the survey, namely that all forms of juvenile cancer as well as leukaemia were more likely to occur when the foetus had been exposed to radiation, was unexpected and to him a little disturbing. On general grounds he would not have expected all tissues to be equally radio-sensitive and to show the same latency period before the appearance of cancer' (FD 1/7826).

8   Experiences of Survivorship

1. Comaroff and Maguire noted that at the time at which they were writing, few studies had been conducted on the psychological impact of coping with the possibility of cure, the only one identified being an earlier paper by O'Malley. The paper in question, entitled 'Long-term Follow-up of Survivors of Childhood Cancer: Psychiatric Sequelae', and presented at the 85th Annual Convention of the American Psychiatric Association in San Francisco in 1977, presents 'a disturbing incidence of psychiatric distress in both the children and their families and raises urgent questions about the

psychological and social implications of survival under improved clinical regimes'.

2. On positive thinking see, among others, Wilkinson and Kitzinger (2000), Rittenberg (1995) and De Raeve (1997). The classic text on cancer and metaphor is Sontag (1979); see also Baines (2009).

3. The clearest version of this story, thoroughly referenced to show its plausibility, is Eiser (2004, Ch. 3). As with paediatric oncology at its time of fastest change – the early 1970s – when most authors of papers and textbooks reflected on their new options and the significance of change, psycho-oncology papers and books from the late 1970s and early 1980s typically opened with the authors' comments about their place and role in (what was presented as) rapid movement in the field.

4. For instance, 1992 saw the launch of the journal *Psycho-oncology*, carrying articles about post-traumatic stress disorder in survivors and families, and personality traits that affected which patients accessed different forms of support. Those within the field in Great Britain have acknowledged the limitations of psycho-oncology's ambitions in the 1970s, when the chief concern was the measurement of learning difficulties in patients treated with cranial radiation.

5. Some practitioners would insist that connections have been made, but none of these supposedly universal characteristics of good copers have been widely accepted – families studied are extremely diverse in their coping strategies, and it has proved hard to show any particular group of subjects to be 'typical' (which would justify the extension of results to the whole population of affected families) even within one country and class, let alone across countries, class, ethnicity, and so on. Many working within psycho-oncology, and those paediatric oncologists who take an interest in it, have been among the first to admit these failings.

6. Thomas Achenbach remains a powerful figure in the field of psychosocial studies, a prominent review writer and synthesiser attempting to bring together diverse sets of terms into one set of agreed-upon definitions and measures, thereby enabling rigorous comparisons (see, e.g., Achenbach, 1995).

7. Interviews with eight paediatric oncologists, conducted in London and Manchester, 2004 and 2005.

8. Treatment centres in Manchester, Liverpool, Southampton, the Royal Marsden, Leeds, and Sheffield have received funding for exactly such a joint long-term project.

9. Parents have a tendency to see their children's upset as anger, while if asked directly children will describe their negative feelings in terms of sadness.

10. We discuss the role played by charities in the provision of social worker support for families later in this chapter.

11. Interview with Pamela Barnes (2004), a long-time trustee of Action for Sick Children.

12. Interviews with paediatric oncologists, conducted in London and Manchester, 2004 and 2005.

13. For comparative histories of American and British child psychiatry, see the work of Marie Reinholdt, Ilana Singh, and Nathan Moon.

14. The peculiar status of cancer and the stigma it carries have been treated in Patterson (1987) and Sontag (1979).

15. This was widely cited over the next few years. See, for example, Spinetta and Deasy-Spinetta (1981) and Koocher and O'Malley (1981).

16. One best-seller was Sutow et al. (1973). This went into a second edition in 1977, in which the preface was amended: 'The demand for a second edition of this specialty textbook so soon after the first is a signal tribute to these advances ... In our first edition, we spoke of a new era of "cautious optimism" in the treatment of children with cancer. There is no question that this phrase is now obsolete, since the cure rate is incontestably improved' (1977, p. x). The advice on psychological support for families nonetheless dealt exclusively with terminal illness. This was the case in every textbook produced in the USA and in Great Britain in the 1970s.

17. We are drawing here on many conversations held by Emm Barnes Johnstone in clinics and wards, and in informal meetings with other 'cancer families' she has met through her son's experience. For a personal and academic reflection on the imperative to 'act normally', offering a fascinating juxtaposition of the perspectives of a sociologist of medicine and her daughter who was treated for cancer as a teenager, see Nancarrow Clarke and Nancarrow Clarke (1999).

18. Sargent was founded in 1967 in Great Britain, from the legacy of conductor Malcolm Sargent. Candlelighters was founded in the USA in 1970 by parents of children with cancer. Cancer and Leukaemia in Childhood (CLIC) began funding play leaders and respite care in 1976, initially in South West England and then nationally. Sargent and CLIC merged in 2005.

19. The history of patient and parent information is traced in Barnes (2006). On charitable funding for psychosocial care initiatives, see Barnes (2005).

20. See also Gerhardt et al. (2007) and Mattsson et al. (2008).

21. See also Bury (2001) and Mathieson and Stam (1995).

References

Archives

The National Archives, Kew, Surrey

Department Code FD: Records created or inherited by the Medical Research Council

FD 1: Medical Research Committee and Medical Research Council: Files

FD 7: Medical Research Council: Committees, Working Parties and Conferences, Registered Files (D Series)

FD 9: Medical Research Council: Registered Files, Policy Matters (A Series)

FD 10: Medical Research Council: Registered Files, Grants (G Series)

FD 23: Medical Research Council: Registered Files, Scientific Matters (S Series)

Department Code MH: Records created or inherited by the Ministry of Health and successors, Local Government Boards and related bodies.

MH 160: Ministry of Health and Department of Health and Social Security: Hospitals, Registered Files (File Office H Series).

John P. McGovern Historical Collections and Research Center, Houston

The Papers of H. Grant Taylor, M.D. Manuscript Collection No. 44.

Series I: Personal and Biographical.

Series II: Atomic Bomb Casualty Commission (ABCC).

Series V: University of Texas M.D. Anderson Hospital and Tumor Institute (now M.D. Anderson Cancer Center).

Papers of Wataru W. Sutow, M.D. Manuscript Collection No. 35.

Series II: Meetings Attended.

Series III: Atomic Bomb Casualty Commission (ABCC).

Medical Sciences Video Archive of The Royal College of Physicians and Oxford Brookes University, Oxford

MSVA: 054, 056, and 059.

Wellcome Library, London

Manuscript collection

WTI/DPB: Burkitt, Denis Parsons (1911–1993).

Cumbrians Opposed to a Radioactive Environment (CORE) archive

Interviews

All interviews were conducted by Emm Barnes Johnstone.
Barnes P. (2004) Manchester.
Chessels J. (2004) London, 22 June 2004.
Craft A. (2004) London, 26 July 2004.
Draper G. and Stiller C. (2004) Oxford, 19 August 2004.
Eden T. (2004) Manchester.
Malpas J. (2004) London, 16 January 2004.
Marsden H. (2005) Manchester, 9 February 2005.
Morris Jones P. (2004) and (2005) London, 22 November 2004 and 31 March 2005.
Pritchard J. (2004) by telephone, 1 August, 2004.

Publications and Websites

Ablett S. (ed.) (2002a) *A History of the UKCCSG Centres* (Leicester: Trident Communications).
Ablett S. (ed.) (2002b) *Quest for Cure: UK Children's Cancer Study Group – The First 25 Years* (Leicester: Trident Communications).
Achenbach T.M. (1966) 'The Classification of Children's Psychiatric Symptoms: A Factor-Analytic Study', *Psychological Monographs* 80:1–37.
Achenbach T.M. (1995) 'Diagnosis, Assessment, and Comorbidity in Psychosocial Treatment Research', *Journal of Abnormal Child Psychology* 23:45–65.
Adams M. (1959) 'Truth About Cancer', *The Times*, 14 Sep.
Aitken-Swan J. and Easson E.C. (1959) 'Reactions of Cancer Patients on Being Told Their Diagnosis', *BMJ* 1(5124):779–83.
Alun Phillips T. (1959) 'Leukaemia and Geography', *The Lancet* 274:659–61.
ApThomas I. (1942) 'Contact X-ray Treatment of Cavernous Angioma in Children', *British Journal of Radiology* 15:43–7.
Arnold D. (ed.) (1988) *Imperial Medicine and Indigenous Societies* (Manchester: Manchester University Press).
Arnold D. (1993) *Colonizing the Body: State Medicine and Epidemic Disease in Nineteenth-Century India* (Berkeley, CA: University of California Press).
Arnold D. (2004) *Science, Technology and Medicine in Colonial India* (Cambridge: Cambridge University Press).
Aronowitz R. (2007) *Unnatural History: Breast Cancer and American Society* (New York: Cambridge University Press).
Austoker, J. (1988) *A History of the Imperial Cancer Research Fund 1902–1986* (Oxford: Oxford University Press).
Austoker J. and Bryder L. (eds) (1989) *Historical Perspectives on the Role of the MRC: Essays in the History of the Medical Research Council of the United Kingdom and its Predecessor. The Medical Research Committee, 1913–1953* (Oxford: Oxford University Press).
Baer R.F. (1955) 'The Sick Child Knows', in Standard S. and Nathan H. (eds) *Should the Patient Know the Truth?*, pp. 100–6 (New York: Springer).

Baines J.E. (2009) *Cancer and the Individual in Britain, 1850 to 2000* Ph.D. University of Manchester.

Baker C. G. (2004) *An Administrative History of the National Cancer Institute's Viruses and Cancer Programs, 1950–1972*, The National Cancer Institute, available at: http://history.nih.gov/research/downloads/SpecialVirusCaPrgm.pdf (accessed 15 September 2014).

Barnes E. (2005) 'Caring and Curing: Paediatric Cancer Services Since 1960', *European Journal of Cancer Care* 14:373–80.

Barnes E. (2006) 'Captain Chemo and Mr Wiggly: Patient Information for Children With Cancer in the Late Twentieth Century', *Social History of Medicine* 19:501–19.

Barnes P. (1999) *Royal Manchester Children's Hospital Pendlebury 1829–1999* (Leek: Churnet Valley Books).

Barnes P. (2010) *Shaping the Future and Celebrating the Past: A Brief History of Sick Children in Greater Manchester 1829–2009* (Manchester: Royal Manchester Children's Hospital).

Bashe C.J., Johnson L.R., Palmer J.H. and Pugh E.W. (1986) *IBM's Early Computers* (Cambridge, MA, and London: MIT Press).

Bell K. and Ristovski-Slijepcevic S. (2013) 'Cancer Survivorship: Why Labels Matter', *Journal of Clinical Oncology* 31:409–11.

Bernard J. (1964) *Recent Research into Human and Experimental Leukaemia* (London: The Queen Anne Press).

Berridge V. (1999) *Health and Society in Britain Since 1939* (Cambridge: Cambridge University Press).

Bird C. (1991) *The Persecution and Trial of Gaston Naessens: The True Story of the Efforts to Suppress an Alternative Treatment for Cancer, AIDS, and Other Immunologically Based Diseases* (Tiburon, CA: H J Kramer Inc.).

Black D. (1984) *Report of the Independent Advisory Group. Investigation of the Possible Increased Incidence of Cancer in West Cumbria* (London: HMSO).

Bluebond-Langner M. (1978) *The Private Worlds of Dying Children* (Princeton, NJ: Princeton University Press).

BMJ (1952) 'Obituary' 1(4765):977.

BMJ (1963) 'Cancer Prevention Detection Centre' 1(5345):277–9.

BMJ (1966) 'Supplement' 2(5513):215.

BMJ (1995) 'Obituary' 311: 1635.

Bond J.V. (1972) 'Unusual Presenting Symptoms in Neuroblastoma', *British Medical Journal* 2(5809):327–8.

Bond J.V. (1975) 'Prognosis and Treatment of Wilms' Tumor at Great Ormond Street Hospital for Sick Children – 1960–1972' *Cancer* 36:1202–7.

Bruce M. (2006) 'A Systematic and Conceptual Review of Posttraumatic Stress in Childhood Cancer Survivors and Their Parents', *Clinical Psychology Review* 26:233–56.

Bryder L. (2011) 'The Medical Research Council and Clinical Trial Methodologies Before the 1940s: The Failure to Develop A "Scientific" Approach', *Journal of the Royal Society of Medicine* 104:335–43.

Bud R. (1978) 'Strategy in American Cancer Research after World War II: A Case Study', *Social Studies of Science* 8:425–59.

Bud R. (2007) *Penicillin: Triumph and Tragedy* (Oxford: Oxford University Press).

Burchenal J.H. (1965) *Approaches to the Aetiology and Treatment of Acute Leukaemia* (London: The Queen Anne Press).

Burchenal J.H. (1966) 'Geographic Chemotherapy – Burkitt's Tumour as a Stalking Horse for Leukemia: Presidential Address', *Cancer Research* 26:2393–405.

Burchenal J.H. (1968) 'Long-term Survivors in Acute Leukemia and Burkitt's Tumor', *Cancer* 21:595–9.

Burchenal J.H. and Murphy M.L. (1965) 'Long-term Survivors in Acute Leukemia', *Cancer Research* 25:1491–5.

Burchenal J.H. and Burkitt D.P. (eds) (1966) *Treatment of Burkitt's Tumour* (Berlin, Heidelberg and New York: Springer-Verlag).

Burkitt D.P. (1958) 'A Sarcoma Involving the Jaws of African Children', *British Journal of Surgery* 46:218–23.

Burkitt D.P. (1962a) 'A "Tumour Safari" in East and Central Africa', *British Journal of Cancer* 16:379–86.

Burkitt D. P. (1962b) 'A Children's Cancer Dependent on Climatic Factors', *Nature* 194:232–4.

Burkitt D.P. (1963) 'A Children's Cancer Related to Climate', *New Scientist* 17:174–6.

Burkitt D. P. (1966) 'Chemotherapy of Jaw Tumours', in Burchenal J.H. and Burkitt D.P. (eds) *Treatment of Burkitt's Tumour*, pp. 94–101 (Berlin, Heidelberg and New York: Springer-Verlag).

Burkitt D.P. (1967) *Possible Connections between the African Lymphoma and Acute Leukaemia* (London: The Queen Anne Press).

Burkitt D.P. and O'Conor G.T. (1961) 'Malignant Lymphoma in African Children. I. A Clinical Syndrome', *Cancer* 14:258–69.

Burkitt D.P. and Wright D.H. (eds) (1970) *Burkitt's Lymphoma* (Edinburgh: E & S Livingstone Ltd).

Bury M. (2001) 'Illness Narratives: Fact or Fiction?', *Sociology of Health and Illness* 23:263–85.

CA: A Cancer Journal for Clinicians (1975) 'Unproven Methods of Cancer Management: Cancer Quackery', *CA: A Cancer Journal for Clinicians* 25:66–71.

Cade S. (1947) 'Discussion on Primary Malignant Tumours of Bone', *British Journal of Radiology* 20:10–18.

Cameron H.C. (1955) *The British Paediatric Association 1928–1952* (London: Metcalfe and Cooper).

Cancer (1975a) 'The Georgetown University Medical Center Cancer Symposium: The Successfully Treated Cancer Patient – New Problems and Challenges January 18, 1974, Washington DC', *Cancer* 36(S1):267–304.

Cancer (1975b) 'The Georgetown University Medical Center Cancer Symposium: The Successfully Treated Cancer Patient – New Problems and Challenges January 18, 1974, Washington DC', *Cancer* 36(S2):623–824.

Cancer Research UK (2012) 'Childhood Cancer Survival Statistics', available at: http://www.cancerresearchuk.org/cancer-info/cancerstats/childhoodcancer/survival/ (accessed 14 November 2012).

Cantor D. (2007) 'Introduction: Cancer Control and Prevention in the Twentieth Century', *Bulletin of the History of Medicine*, 81:1–38.

Cantor D. (ed.) (2008) *Cancer in the Twentieth Century* (Baltimore, MD, and London: Johns Hopkins University Press).

Cantrell M.A. and Conte T.M. (2009) 'Between Being Cured and Being Healed: The Paradox of Childhood Cancer Survivorship', *Qualitative Health Research* 19:312–22.

Cantwell A.R. (2003) 'The Russell Body – The Forgotten Clue to the Bacterial Cause of Cancer', *Journal of Independent Medical Research* 1:1.

Cherry S. (1996) *Medical Services and the Hospitals in Britain, 1860–1939* (Cambridge: Cambridge University Press).

Children's Cancer and Leukaemia Group (2012) 'About Children's Cancer and Leukaemia Group', available at: www.cclg.org.uk/about-us (accessed 10 December 2012).

Christie D.A. and Tansey E.M. (eds) (2003) *Wellcome Witness To Twentieth Century Medicine Volume 15: Leukaemia* (London: The Wellcome Trust).

Cochran A. J. (1978) *Man, Cancer and Immunity* (London: Academic Press).

Comaroff J. and Maguire P. (1981) 'Ambiguity and the Search for Meaning: Childhood leukaemia in the Modern Clinical Context', *Social Science and Medicine* 15B:115–23.

Cooter R. (1992) *In the Name of the Child: Health and Welfare, 1880–1940* (London: Routledge).

Cooter R. and Pickstone J.V. (eds) (2003) *Companion to Medicine in the Twentieth Century* (London: Routledge).

Court Brown W.M. and Doll R. (1957) *Leukaemia and Aplastic Anaemia in Patients Irradiated for Ankylosing Spondylitis, MRC Special Report Series, No. 295* (London: HMSO).

Coward R. (1989) *The Whole Truth. The Myth of Alternative Health* (London: Faber and Faber).

Creager A.N.H. (2001) *The Life of A Virus: Tobacco Mosaic Virus as an Experimental Model, 1930–1965* (Chicago, IL: The University of Chicago Press).

Creager A.N.H. and Gaudillière J.-P. (2001) 'Experimental Arrangements and Technologies of Visualization: Cancer as a Viral Epidemic (1930–1960)', in Gaudillière J.-P. and Löwy I. (eds) *Heredity and Infection: The History of Disease Transmission*, pp. 203–41 (London: Routledge).

Croarken M. (1990) *Early Scientific Computing in Britain* (Oxford: Clarendon Press).

Curtis F. (1959) 'Truth About Cancer' *The Times*, 16 Sep.

Cutler J. (1985) 'Close to the Bone: James Cutler Reveals How the Black Report on Child Cancer Around Windscale is to be Followed up', *New Statesman* 1 Feb.

D'Angio G.J., Beckwith J.B., Breslow N.E., Bishop H.C., Evans A.E., Farewell V., et al. (1980) 'Wilm's Tumor: An Update', *Cancer* 45(7 Suppl.):1791–8.

Dargeon H. (ed.) (1940) *Cancer in Childhood. And a Discussion of Certain Benign Tumors* (St Louis, MO: C. V. Mosby Company).

Davis D. (2007) *The Secret History of the War on Cancer* (New York: Perseus).

De Raeve L. (1997) 'Positive Thinking and Moral Oppression in Cancer Care', *European Journal of Cancer Care* 6:249–56.

Derek R., Jenkin T. and Berry M.P. (1980) 'Hodgkin's Disease in Children', *Seminars in Oncology* 7:202–11.

DeVita V. and Chu E. (2008) 'A History of Cancer Chemotherapy', *Cancer Research* 68:8643–53.

DeVita V., Lewis B.J., Rozencweig M. and Muggia F.M. (1978) 'The Chemotherapy of Hodgkin's Disease: Past Experiences and Future Directions', *Cancer* 42:979–90.

Dietz A.C. and Mulrooney D.A. (2011) 'Life Beyond the Disease: Relationships, Parenting, and Quality of Life Among Survivors of Childhood Cancer', *Haematologica* 96:643–5.

Dobbs R.H. (1953) 'Antibiotic and Chemotherapeutic Agents in the Treatment of Infantile Diarrhea and Vomiting', *The Lancet* 265:1163–9.

Drew S. (2007) '"Having Cancer Changed my Life, and Changed my Life Forever": Survival, Illness Legacy and Service Provision Following Cancer in Childhood', *Chronic Illness* 3:278–95.

Dry S. (2004) 'The Population as Patient', in Schich T. and Tröhler U. (eds) *The Risks of Medical Innovation: Risk Perception and Assessment in Historical Context*, pp. 116–32 (London: Routledge).

Dunn P.M. (2006) 'Sir Frederick Still (1868–1941): The Father of British Paediatrics', *Archives of Disease in Childhood, Fetal and Neonatal Edition* 91:308–10.

Easson E.C. (1966a) 'Long Term Results of Radical Radiotherapy in Hodgkin's Disease', *Cancer Research* 26:1244.

Easson E.C. (1966b) 'Radiotherapy for Malignant Lymphomas', *Annual Review of Medicine* 17:179–96.

Edwards, M. (2007) *Control and the Therapeutic Trial: Rhetoric and Experimentation in Britain, 1918–1948* (Amsterdam: Editions Rodopi B.V.)

Eiser C. (2004) *Cancer in Children* (London: Lawrence Erlbaum).

Epstein A. (2004) 'Burkitt, Denis Parsons (1911–1993)', in *Oxford Dictionary of National Biography* (Oxford: Oxford University Press).

Evans A.E. (1968) 'If a Child Must Die', *New England Journal of Medicine* 278:138–42.

Falcone R. (2005) 'The Cancer Bacteria Homepage' available at: http://members. aol.com/CAbacteria/homepage.html (accessed 12 January 2006).

Farber S. (1965) 'Management of the acute leukemia patient and Family', *CA: A Cancer Journal for Clinicians* 15:14–17.

Farber S., Diamond L.K., Mercer R.D., Sylvester R.F. and Woolf J.A. (1948) 'Temporary Remissions in Acute Leukemia in Children Produced by Folic Acid Antagonist, 4-Aminopteroyl-glutamic acid (Aminopterin)', *New England Journal of Medicine* 238:787–93.

Feudtner J.C. (2003) *Bittersweet: Diabetes, Insulin, and the Transformation of Illness* (Chapel Hill, NC, and London: University of North Carolina Press).

Francis L. (1959) 'Truth About Cancer' *The Times*, 16 Sep.

Frei E. and Freireich E. J. (1965) 'Progress and Perspectives in the Chemotherapy of Acute Leukemia', *Advances in Chemotherapy* 2:269–98.

Freireich E.J., Karon M. and Frei, E. (1964) 'Quadruple Combination Therapy (VAMP) for Acute Lymphocytic Leukemia in Childhood', *Proceedings of the American Association of Cancer Research* 5:20.

Fuller L.M., Sullivan M.P. and Butler J.J. (1973) 'Results of Regional Radiotherapy in Localized Hodgkin's Disease in Children', *Cancer* 32:640–5.

Furnham A. and Smith C. (1988) 'Choosing Alternative Medicine. A Comparison of the Beliefs of Patients Visiting a General Practitioner and a Homeopath', *Social Science and Medicine* 26:685–9.

Gale A.H., Gillespie W.A. and Perry C.B. (1952) 'Oral Penicillin in the Prophylaxis of Streptococcal Infection in Rheumatic Children', *The Lancet* 2:61–3.

Gardner M. (1993) 'Review of Reported Increases of Childhood Cancer Rates in the Vicinity of Nuclear Installations in the United Kingdom', in Beral V. and Roman E. (eds) *Childhood Cancer and Nuclear Installations: Papers, Abstracts, Letters, Editorials, Reports Published Since 1984*, pp. 146–66 (London: British Medical Journal).

Gaudillière J.-P. (1999) 'Circulating Mice and Viruses: The Jackson Memorial Laboratory, the National Cancer Institute, and the Genetics of Breast Cancer, 1930–1965', in Fortun M. and Mendelsohn E. (eds) *Practices of Human Genetics*, pp. 89–124 (Dordrecht: Kluwer Academic Publishers).

Gaudillière J.-P. (2006) 'The Cancer Century', in Bowler P.J. and Pickstone J.V. (eds) *The Cambridge History of Science, Vol. 6: The Biomedical and Earth Sciences since 1800 The Modern Biological and Earth Sciences* (Cambridge: Cambridge University Press).

George S.L., Rhomes J.A. Aur, Mauer A.M. and Simone J.V. (1979) 'A Reappraisal of the Results of Stopping Therapy in Childhood Leukemia', *New England Journal of Medicine* 300:269–73.

Gerhardt C.A., Vannatta K., Valerius K.S., Correll J. and Noll R.B. (2007) 'Social and Romantic Outcomes in Emerging Adulthood Among Survivors of Childhood Cancer', *Journal of Adolescent Health* 40:462.e9–462.e15.

Giles D., Hewitt D., Stewart A. and Webb J. (1956) 'Malignant Disease in Childhood and Diagnostic Irradiation in Utero', *The Lancet* 271:447.

Glemser B. (1971) *The Long Safari* (London: Bodley Head).

Goldman J.M. and Gordon M.Y. (2003) 'A History of the Chronic Leukemias' in Wiernik P.H., Goldman J.M., Dutcher J.P. and Kyle R.A. (eds) *Neoplastic Diseases of the Blood*, 4th ed., pp. 3–8 (Cambridge: Cambridge University Press).

Gould D. (1964a) 'Is Leukaemia Caused by a Virus?', *New Scientist* 21:402–3.

Gould D. (1964b) 'Viruses Linked With Leukaemia', *New Scientist* 21:142–3.

Great Britain Advisory Committee on Cancer Registration (1981) *Report of the Advisory Committee on Cancer Registration: Cancer Registration in the 1980s* (London: HSMO).

Greene G. (1999) *The Woman Who Knew Too Much: Alice Stewart and the Secrets of Radiation* (Ann Arbor, MI: The University of Michigan Press).

Greene G. (2011) 'Richard Doll and Alice Stewart: Reputation and the Shaping of "Scientific Truth"', *Perspectives in Biology and Medicine* 54:505–32.

Greer H. S. (1981) 'Psychobiological Studies of Women With Breast Cancer', *Journal of Psychosomatic Research* 25:450.

Greer S., Morris T. and Pettingale K.W. (1979) 'Psychological Response to Breast Cancer: Effect on Outcome', *The Lancet* 314:782–7.

Greulich W.W. and Pyle S. I. (1950) *Radiographic Atlas of Skeletal Development of the Wrist and Hand* (Stanford, CA: Stanford University Press).

Haddow A. (1961) 'Research Into Cancer: Progress That has Been Made' *The Times*, 16 May.

Hall C. (ed.) (2000) *Cultures of Empire: A Reader* (Manchester: Manchester University Press).

Hall S. (ed.) (1997) *Representation: Cultural Representation and Signifying Practices* (London: Sage).

Haraway D. (1993) 'The Biopolitics of Postmodern Bodies. Determinations of Self in Immune System Discourse' in Lindenbaum S. and Lock M. (eds) *Knowledge, Power and Practice. The Anthropology of Medicine in Everyday Life*, pp. 364–410 (Berkley, CA: University of California Press).

Harrison M. (1994) *Public Health in British India: Anglo-Indian Preventive Medicine 1859–1914* (Cambridge: Cambridge University Press).

Hayhoe F.G.J. (ed.) (1965) *Current Research in Leukaemia* (Cambridge: Cambridge University Press).

Helfand W.H., Lazarus J. and Theerman P. (2001) '"... So That Others May Walk": The March of Dimes', *American Journal of Public Health* 91:1190.

Hewitt D. (1955) 'Some Features of Leukaemia Mortality', *British Journal of Preventive and Social Medicine* 9:81–8.

Higgs E. (2000) 'Medical Statistics, Patronage and the State: The Development of the MRC Statistical Unit, 1911–1948', *Medical History* 44:323–40.

Historical Archives Advisory Committee (2001) 'Committee Report: American Pediatrics: Milestones at the Millenium', *Pediatrics* 107:1482–91.

Holland J.C. (1982) 'Why Patients Seek Unproven Cancer Remedies: A Psychological Perspective', *CA: A Cancer Journal for Clinicians* 32:10–14.

Howard H.I. (1959) 'Truth About Cancer' *The Times*, 16 Sep.

Howard S. (1959) 'Truth About Cancer' *The Times*, 15 Sep.

Hustu H.O., Aur R.J., Verzosa M.S., Simone J.V. and Pinkel D. (1973) 'Prevention of Central Nervous System Leukemia by Radiation', *Cancer* 32:585–97.

Jackson A.D.M. (1988) 'The British Paediatric Association: 60 years', *Archives of Disease in Childhood* 63:229–30.

James K.W. and Kay H.E.M. (1967) 'Some Aspects of Current Therapy in Acute Leukaemia', *The Lancet* 289:206–9.

Javier R.T. and Butel J.S. (2008) 'The History of Tumor Virology', *Cancer Research* 68:7693–706.

Jenkins R.G.C. (1959) 'Truth About Cancer' *The Times*, 12 Sep.

Jolliffe D.M. (1993) 'A History of the use of Arsenicals in Man', *Journal of the Royal Society of Medicine* 86:287–9.

Keating P. and Cambrosio A. (2002) 'From Screening to Clinical Research: The Cure of Leukemia and the Early Development of the Cooperative Oncology Groups, 1955–1966', *Bulletin of the History of Medicine* 76:29–34.

Keating P. and Cambrosio A. (2007) 'Cancer Clinical Trials: The Emergence and Development of a New Style of Practice', *Bulletin of the History of Medicine* 81:187–223.

Keating P. and Cambrosio A. (2012) *Cancer on Trial: Oncology as a New Style of Practice* (Chicago, IL: University of Chicago Press).

Kellock B. (1985) *The Fibre Man: The Life Story of Dr Denis Burkitt* (Tring: Lion Publishing).

Khan N.F., Harrison S., Rose P.W., Ward A. and Evans J. (2012) 'Interpretation and Acceptance of the Term "Cancer Survivor": A United Kingdom-based Qualitative Study', *European Journal of Cancer Care* 21:177–86.

Klawiter M. (2004) 'Breast Cancer in Two Regimes: The Impact of Social Movements on Illness Experience', *Sociology of Health and Illness* 26:845–74.

Kleinman A. (1988) *Illness Narratives. Suffering, Healing, and the Human Condition* (New York: Basic Books).

Koocher G.P. and O'Malley J.E. (eds) (1981) *The Damocles Syndrome: Psychosocial Consequences of Surviving Childhood Cancer* (London: McGraw-Hill).

Kneale, G.W. (1971) 'Excess Sensitivity of Pre-leukaemics to Pneumonia: A Model Situation for Studying the Interaction of Infectious Disease With Cancer', *British Journal of Preventive and Social Medicine* 25:152–9.

Krueger G.M. (2008) *Hope and Suffering: Children, Cancer, and the Paradox of Experimental Medicine* (Baltimore, MD: Johns Hopkins University Press).

Lancashire E.R., Frobisher C., Reulen R.C., Winter D.L., Glaser A. and Hawkins, M.M. (2010) 'Educational Attainment Among Adult Survivors of Childhood

Cancer in Great Britain: A Population-based Cohort Study', *Journal of the National Cancer Institute* 102:254–70.

Laszlo J. (1995) *The Cure of Childhood Leukaemia: Into the Age of Miracles* (New Brunswick, NJ: Rutgers University Press).

Latour B. (1987) *Science in Action: How to Follow Scientists and Engineers Through Society* (Milton Keynes: Open University Press).

Latour B. and Woolgar S. (1979) *Laboratory Life: the Social Construction of Scientific Facts* (London: Sage).

Lauder I. and Aherne W. (1972) 'The Significance of Lymphocytic Infiltration in Neuroblastoma', *British Journal of Cancer* 26:321–30.

Law L.W. (1952) 'Effects of Combinations of Antileukemic Agents on an Acute Lymphocytic Leukemia of Mice', *Cancer Research* 72:871–8.

LeCount E.R. (1902) 'The Analogies Between Plimmer's Bodies and Certain Structures Found Normally in the Cytoplasm', *Journal of Medical Research* 7:383–93.

Lederman M. and Jones C.H. (1974) *Malignant Diseases in Children* (London: Butterworth).

Le Fanu L. (1999) *The Rise and Fall of Modern Medicine* (London: Little, Brown and Company).

Lennox E.L., Stiller C.A., Morris Jones P.H. and Kinnier Wilson L.M. (1979) 'Nephroblastoma: treatment during 1970–3 and the effect on survival of inclusion in the first MRC trial', *British Medical Journal* 2(6190):567–9.

Leonard (1984) 'Windscale Gets Black Whitewash' *The Irish Independent* 24 Jul.

Lerner B.H. (2001) *The Breast Cancer Wars: Hope, Fear, and the Pursuit of a Cure in Twentieth–Century America* (New York: Oxford University Press).

Lewis J. (1992) 'Providers, Consumers, the State and the Delivery of Health-care Services in Twentieth-century Britain' in Wear A. (ed.) *Medicine in Society: Historical Essays*, pp. 317–45 (Cambridge: Cambridge University Press).

Lightwood R., Barrie H. and Butler N. (1960) 'Observations on 100 Cases of Leukaemia in Childhood', *British Medical Journal* 1(747):747–52.

Lilleyman, J. (2003) 'Leukaemia', in Christie D.A. and Tansey E.M. (eds) (2003) *Wellcome Witness To Twentieth Century Medicine Volume 15: Leukaemia*, pp. 15–84 (London: The Wellcome Trust).

Lindee M. S. (1994a) 'Atonement: Understanding the No-treatment Policy of the Atomic Bomb Casualty Commission', *Bulletin of the History of Medicine* 68(3):454–90.

Lindee M.S. (1994b) *Suffering Made Real: American Science and the Survivors of Hiroshima* (London: University of Chicago Press).

Little M., Jordens C.F.C., Montgomery P.K. and Philipson B. (1998) 'Liminality: A Major Category of the Experience of Cancer Illness', *Social Science and Medicine* 47:1485–94.

Livingstone J. (2012) *Improvising Medicine: An African Oncology Ward in an Emerging Cancer Epidemic* (Durham, NC: Duke University Press).

Lomax E. (1996) 'Small and Special: The Development of Hospitals for Children in Victorian Britain', *Medical History Supplement* 16:1–217.

Löwy I. (1994) 'Experimental Systems and Clinical Practices: Tumor Immunology and Cancer Immunotherapy, 1895–1980', *Journal of the History of Biology* 27:403–35.

Löwy I. (1996) *Between Bench and Bedside: Science, Healing, and Interleukin-2 in a Cancer Ward* (Cambridge, MA, and London: Harvard University Press).

Löwy I. (2001) 'Images of the New Cancer Therapies' in Löwy I. and Krige J. (eds) *Images of Disease. Science, Public Policy and Health in Post-War Europe*, pp. 73–87 (Luxembourg: Official Publications of the European Communities).

McIntosh J. (1976) 'Patients' Awareness and Desire for Information About Diagnosed but Undisclosed Malignant Disease', *The Lancet* 308:300–3.

Magnello E. and Hardy A. (eds) (2002) *The Road to Medical Statistics* (Amsterdam: Rodopi BV).

Malpas J.S. and Freeman J.E. (1974) 'Blood and Neoplastic Diseases: Solid Tumours in Children', *British Medical Journal* 4(5946):710–13.

Malpas J.S., Freeman J.E., Paxton A., Walker Smith J., Stansfeld A.G. and Wood C.B.S. (1976) 'Radiotherapy and Adjuvant Combination Chemotherapy for Childhood Rhabdomyosarcoma', *BMJ* 1(6004):247–9

Marks H. (1998) *The Progress of Experiment: Science and Therapeutic Reform in the United States, 1900–1990* (Cambridge: Cambridge University Press).

Marsden H.B. and Steward J.K. (eds) (1968a) *Tumours in Children* (Berlin: Springer).

Marsden H.B. and Steward J.K. (1968b) 'Cancer in Children', *The Lancet* 292:733.

Martin E. (1994) *Flexible Bodies. Tracking Immunity in American Culture – From the Days of Polio to the Age of AIDs* (Boston, MA: Beacon Press).

Mathieson C.M. and Stam H.J. (1995) 'Renegotiating Identity: Cancer Narratives', *Sociology of Health and Illness* 17:283–306.

Mattsson E., Lindgren B. and von Essen L. (2008) 'Are There any Positive Consequences of Childhood Cancer? A Review of the Literature', *Acta Oncologica* 47:199–206.

M.D. Anderson Hospital and Tumor Institute (1965) *Historical Highlights of The University of Texas M. D. Anderson Hospital and Tumor Institute* (Houston, TX: The University of Texas M. D. Anderson Hospital and Tumor Institute).

Medical Research Council (1956) *The Hazards to Man of Nuclear and Allied Radiation* (London: HMSO).

Medical Research Council (1971) 'Duration of Survival of Children With Acute Leukaemia: Report to the Medical Research Council from the Committee on Leukaemia and the Working Party on Leukaemia in Childhood', *British Medical Journal* 4(5778):7–9.

Medical Research Council's Working Party on Embryonal Tumours in Childhood (1978) 'Management of Nephroblastoma in Childhood. Clinical Study of Two Forms of Maintenance Chemotherapy', *Archives of Disease in Childhood* 53:112–19.

Michel G., Greenfield D.M., Absolom K., Ross R.J., Davies H. and Eiser C. (2009) 'Follow-up Care After Childhood Cancer: Survivors' Expectations and Preferences for Care', *European Journal of Cancer* 45:1616–23.

Miller R.W. (1968) 'Effects of Ionizing Radiation From the Atomic Bomb on Japanese Children', *Pediatrics* 41:257–63.

Moscucci O. (1993) *The Science of Woman: Gynaecology and Gender in England, 1800–1929* (Cambridge: Cambridge University Press).

Moscucci O., Herring R. and Berridge V. (2009) 'Networking Health Research in Britain: The Post-war Childhood Leukaemia Trials', *Twentieth Century British History* 20:23–52.

Mott M.G., Mann J.R. and Stiller C.A. (1997) 'The United Kingdom Children's Cancer Study Group – the First 20 Years of Growth and Development', *European Journal of Cancer* 33:1448–52.

Moulin A.M. (1989) 'The Immune System: A Key Concept for the History of Immunology', *History and Philosophy of the Life Sciences* 11:221–36.

Murphy C.C.S. (1986) 'A History of Radiotherapy to 1950. Cancer and Radiotherapy 1850–1950', unpublished PhD thesis, University of Manchester.

Murray J. and Shepherd S. (1993) 'Alternative or Additional Medicine? An Exploratory Study in General Practice', *Social Science and Medicine* 37:983–8.

Nancarrow Clarke J. and Nancarrow Clarke L. (1999) *Finding Strength: A Mother and Daughter's Story of Childhood Cancer* (Toronto: Oxford University Press).

National Cancer Institute (1969) *Report of the Williamsburg Conference, October 20–23, 1968: Recommendations of the Cancer Clinical Investigation Review Committee and the National Advisory Cancer Council regarding the Cooperative Clinical Cancer Research Program* (Bethesda, MD: National Cancer Institute).

Nature (1964) 'Funds for Cancer Research', 201:23.

Nicolas V. (1959) 'Truth About Cancer' *The Times*, 18 Sep.

O'Conor G.T. (1961) 'Malignant Lymphoma in African Children. II. A Pathological Entity', *Cancer* 14:270–83.

Olson J.S. (2009) *Making Cancer History: Disease and Discovery at The University of Texas M. D. Anderson Cancer Center* (Baltimore, MD: Johns Hopkins University Press).

Paice E. (1999) 'History: The History of Safari', *Travel Africa* Edition 10, available at: http://www.travelafricamag.com/index2.php?option=com_content&do_pdf=1&id=128 (accessed 20 November 2012).

Pantin A.M. (1959) 'Truth About Cancer' *The Times*, 15 Sep.

Parascandola M. (2001) 'Cigarettes and the US Public Health Service in the 1950s', *American Journal of Public Health* 91:196–205.

Paterson E. (1955) 'Malignant Disease in Children', *South African Medical Journal* 29:1199–206.

Paterson R. and Aitken-Swann J. (1954) 'Public Opinion on Cancer. A Survey Among Women in the Manchester Area', *The Lancet* 264:857–61.

Patterson J.T. (1987) *The Dread Disease: Cancer and Modern American Culture* (Cambridge, MA, and London: Harvard University Press).

Patterson J.T. (1991) 'Cancer, Cancerphobia, and Culture. Reflections on Attitudes in the United States and Great Britain', *Twentieth Century British History* 2:137–42.

Pearson D. (1964) 'The Role of Radiotherapy in the Treatment of Tumours in Children', *Journal of Clinical Pathology* 17:423–6.

Pearson D., Duncan W.B. and Pointon R.C.S. (1964) 'V. Wilm's Tumours – A Review of 96 Consecutive Cases' *British Journal of Radiology* 37:154–60.

Penson R.T., Schapira L., Daniels K.J., Chabner B.A. and Lynch T.J. (2004) 'Cancer as Metaphor', *The Oncologist* 9:708–16.

Pettingale K.W. (1984) 'Coping and Cancer Process', *Journal of Psychosomatic Research* 28:363–4.

Pettingale K.W., Morris T., Greer S. and Haybittle J.L. (1985) 'Mental Attitudes to Cancer: An Additional Prognostic Factor', *The Lancet* 325:750.

Pickstone J.V. (2003) 'Production, Community and Consumption: The Political Economy of Twentieth-century Medicine', in Cooter R. and Pickstone J.V.

(eds) *Companion to Medicine in the Twentieth Century*, pp. 1–19 (London: Routledge).

Pickstone J.V. (2007) 'Contested Cumulations: Configurations of Cancer Treatments Through the Twentieth Century', *Bulletin of the History of Medicine* 81:164–96.

Piller G.J. (1994) *Rays of Hope: The Story of the Leukaemia Research Fund* (London: The Leukaemia Research Fund).

Piller G.J. (2001) 'Leukaemia: A Historical Review From Ancient Times to 1950', *British Journal of Haematology* 112:282–92.

Pinell P. (2002) *The Fight Against Cancer: France 1890–1940* (London: Routledge).

Pinell P. (2003) 'Cancer', in Cooter R. and Pickstone J.V. (eds) *Companion to Medicine in the Twentieth Century*, pp. 671–86 (London: Routledge).

Pinkel D. (1972) 'Nine Years' Experience With "Total Therapy" of Childhood Acute Lymphocytic Leukemia', *Pediatrics* 50:246–51.

Pinkel D. (1985) 'Current Issues in the Management of Children With Acute Lymphocytic Leukaemia', *Postgraduate Medical Journal* 61:93–102.

Quirke V. and Gaudillière J.-P. (2008) 'The Era of Biomedicine: Science, Medicine, and Public Health in Britain and France After the Second World War', *Medical History* 52:441–52.

Radford E. (1983) 'The Windscale Controversy' *Financial Times*, 17 Nov.

Rasmussen N. (1997) *Picture Control: The Electron Microscope and the Transformation of Biology in America, 1940–1960* (Stanford, CA: Stanford University Press).

Renouf J. (1993) *Jimmy. No Time to Die* (London: Fontana).

Rhomes J.A.A., Simone J.V., Omar Hustu H. and Verzosa M.S. (1972) 'A Comparative Study of Central Nervous System Irradiation and Intensive Chemotherapy Early in Remission of Childhood Acute Lymphocytic Leukemia', *Cancer* 29:381–91.

Rimmon-Kenan S. (2002) 'The Story of "I": Illness and Narrative Identity', *Narrative* 10:9–27.

Rittenberg C. (1995) 'Positive Thinking: An Unfair Burden for Cancer Patients?', *Supportive Care in Cancer* 3:37–8.

Rosser Matthews J. (1995) *Quantification and the Quest for Medical Certainty* (Princeton, NJ: Princeton University Press).

Rossi P.N. (2009) *Fighting Cancer With More Than Medicine: A History of Macmillan Cancer Support* (Stroud: The History Press).

Rous P. (1910) 'A Transmissable Avian Neoplasm (Sarcoma of the Common Fowl)', *Journal of Experimental Medicine* 12:696–705.

Rous P. (1911) 'A Sarcoma of the Fowl Transmissible by an Agent Separable From the Tumor Cells', *Journal of Experimental Medicine* 13:397–411.

Rouse A. (1959) 'Truth About Cancer' *The Times*, 15 Sep.

Russell W. (1890) 'An Address on a Characteristic Organism of Cancer', *BMJ* 2:1356–60.

Seager J. (2003) 'Rachel Carson Died of Breast Cancer', *Journal of Women in Culture and Society* 28:945–72.

Shapiro D.M. and Gellhorn A. (1951) 'Combinations of Chemical Compounds in Experimental Cancer Therapy', *Cancer Research* 11:35–41.

Sharma U. (1992) *Complementary Medicine Today. Practitioners and Patients* (London: Routledge).

Shulman S. T. (2004) 'A History of Pediatric Infectious Diseases', *Pediatric Research* 55:163–76.

Sinclair E.A. (1959) 'Truth About Cancer' *The Times*, 18 Sep.

Smith E. (1964) *To The Bitter End* (London: Abelard-Schuman).

Smith I.E., Peckham M.J., McElwain T.J., Garret J.C. and Austin D.E. (1977) 'Hodgkin's Disease in Children', *British Journal of Cancer* 36:120–9.

Smyth R. (1964) 'No More Pleads Serum Man' *The Daily Mail*, 4 Jan.

Sontag S. (1979) *Illness as Metaphor* (London: Allen Lane).

Spackman A. and Connett D. (1986) 'New Sellafield Scandal: Government Admits True Level of Radiation was Concealed for 30 Years', *The Sunday Times*, 16 Feb.

Spiers A.S.D. (1972) 'Cure as the Aim in Therapy for the Acute Leukaemias', *The Lancet* 300:473–5.

Spinetta J.J. and Deasy-Spinetta P. (eds) (1981) *Living with Childhood Cancer* (London: The C. V. Mosby Company).

Standard S. and Nathan H. (1955) 'Preface', in Standard S. and Nathan H. (eds) *Should the Patient Know the Truth?*, pp. 9–10 (New York: Springer).

Starr P. (1982) *Transformation of American Medicine: The Rise of a Sovereign Profession and the Making of a Vast Industry* (New York: Basic Books).

Stevens R.F. (2001) 'Sir Leonard Parsons and the Scientific Basis of Paediatric Haematology', *British Journal of Haematology* 112:558–560.

Stewart A.M. and Kneale G.W. (1969) 'The Role of Local Infections in the Recognition of Haemopoietic Neoplasms', *Nature* 223:741–2.

Stewart A. and Kneale G.W. (1970a) 'Prenatal Radiation Exposure and Childhood Cancer', *The Lancet* 296:1189–90.

Stewart A.M. and Ledlie E.M. (1968) 'Cancer in Children', *The Lancet* 292:453–4.

Stiller, C.A. (1994) 'Population Based Survival Rates for Childhood Cancer in Britain, 1980–91'. *BMJ* 309:1612.

Stiller C.A. (2007) *Childhood Cancer in Britain: Incidence, Survival, Mortality* (Oxford: Oxford University Press).

Sutow W. (1965) 'Chemotherapy in Childhood Cancer (Except Leukemia): An Appraisal', *Cancer* 18:1585–9.

Sutow W., Vietti T.J. and Fernbach D.J. (eds) (1973) *Clinical Pediatric Oncology* (St Louis, MO: The C. V. Mosby Company).

Sutow W., Vietti T.J. and Fernbach D.J. (eds) (1977) *Clinical Pediatric Oncology*, 2nd ed. (St. Louis, MO: The C. V. Mosby Company).

Tan C.T.C., Dargeon H.W. and Burchenal J.H. (1959) 'The Effect of Actinomycin D on Cancer in Childhood', *Pediatrics* 24:544–61.

Tauber A.I. (1994) *The Immune Self: Theory or Metaphor?* (Cambridge: Cambridge University Press).

Taylor G. (ed.) (1990) *Pioneers in Pediatric Oncology* (Austin, TX: University of Texas Press).

Telfer K. (2008) *The Remarkable Story of Great Ormond Street Hospital* (London: Simon and Schuster).

The Children's Solid Tumour Group (1981) 'The Long-term Outlook for Children Treated for Non-Hodgkin's Lymphomas', *British Journal of Cancer* 44:872–8.

The Daily Telegraph (1963) 'Leukaemia Boy Gets Serum. Treatment Begins in Corsica Clinic', 28 Dec.

The Daily Telegraph (1965) 'Naessens Gets Maximum Fine', 15 May.

The Daily Telegraph (1984) 'Doctor Angry Over Cancer Risk Idea', 16 Oct.

The Guardian (1959) 'Association Between Fall-out and Leukaemia Deaths', 23 Oct.

The Guardian (1984) 'Lingering Particles of Unease', 24 Jul.

The Lancet (1965) 'Acute Leukaemia', 286:938–39.

The Lancet (1968) 'Cancer in Children', 292:32–3.

The Lancet (1971) 'Long Remissions in Leukaemia', 297:742–3.

The Times (1959a) 'Surgeon's Plea on Cancer Victims: "Should not be Told"', 10 Sep.

The Times (1959b) 'Surgeon Defends Shielding Patients From Truth' 13 Oct.

The Times (1960) 'Control of a Rare Form of Cancer', 6 Jan.

The Times (1961a) 'Research Unit for Leukaemia', 7 Dec.

The Times (1961b) 'U.S. Grant for Cancer Fight: Oxford Hope of Early Result', 23 Jan.

The Times (1961c) 'Intense Cancer Research', 28 Jun.

The Times (1962) 'Is There a Virus Origin for Cancer?', 27 Jul.

The Times (1963a) 'Virus Link With Tumours Sought', 10 Apr.

The Times (1963b) 'Virus Link With Malignant Disease', 24 Jul.

The Times (1964a) 'A Mystery Virus Detected in Leukaemia', 18 Sep.

The Times (1964b) 'Cancer Treatment Unit is "Landmark in Medicine": Rare Form Yields to Course of Drugs', 4 Mar.

The Times (1964c) 'Success Claimed for Cancer Unit', 16 Oct.

The Times (1964d) 'A Cancer Research Question Mark', 25 Nov.

The Times (1964e) 'French "Cancer Serum" to be Tested', 8 Jan.

The Times (1964f) 'M. Naessens Coming to Britain', 13 Jan.

The Times (1965a) 'Trial of "Leukaemia Cure" Man Opens in Paris', 4 May.

The Times (1965b) 'Experts Give Evidence on Leukaemia', 5 May.

The Times (1965c) 'Maximum Fine of £1,300 Imposed on Naessens: Cancer Cure Disproved, "Typical Tinkerer"', 15 May.

The Times (1973) 'Medicine: Study of Childhood Leukaemia', 10 Mar.

The Practitioner (1899) *Cancer Issue* New Series IX.

The Whitehaven News (1985) 'Plutonium Tests to Start on Foetuses', 7 Feb.

Thomas K.J., Carr J., Westlake L. and Williams B.T. (1991) 'Use of Non-orthodox and Conventional Health Care in Great Britain', *BMJ* 302:207–10.

Thompson R.B. and Walker W. (1962) 'Study of 50 Cases of Acute Leukaemia in Childhood', *British Medical Journal* 1(5286):1165–69.

Thomson A.L. (1974) *Half a Century of Medical Research: Volume One: Origins and Policy of the Medical Research Council* (London: HMSO).

Thomson A.L. (1976) *Half a Century of Medical Research: Volume Two: The Programme of the Medical Research Council* (London: HMSO).

Timmermann C. (2008) 'As Depressing as it was Predictable? Lung Cancer, Clinical Trials, and the Medical Research Council in Postwar Britain', in Cantor D. (ed.) *Cancer in the Twentieth Century*, pp. 312–34 (Baltimore, MD: Johns Hopkins University Press).

Timmermann C. (2012) '"Just Give me the Best Quality of Life Questionnaire": The Karnofsky Scale and the History of Quality of Life Measurements in Cancer Trials', *Chronic Illness* 9:179–90.

Timmermans S. and Berg M. (2003) *The Gold Standard: The Challenge of Evidence-Based Medicine and Standardization in Health Care* (Philadelphia, PA: Temple University Press).

Tod M.C. (1940) 'The Treatment of Metastases', *British Journal of Radiology* 13:163–71.

Toon E. (2007) '"Cancer as the General Population Knows it": Knowledge, Fear, and Lay Education in 1950s Britain', *Bulletin of the History of Medicine*, 81:116–38.

Valier H. and Timmermann C. (2008) 'Clinical Trials and the Reorganization of Medical Research in Post-Second World War Britain', *Medical History* 52:493–510.

Valman B. (ed.) (2000) *The Royal College of Paediatrics and Child Health at the Millennium* (London: Royal College of Paediatrics and Child Health).

Van Epps H. L. (2005) 'Peyton Rous: Father of the Tumor Virus', *Journal of Experimental Medicine* 201:320.

van Eys J. (1977a) 'The Outlook for the Child With Cancer', *Journal of School Health* 47:165–9.

van Eys J. (ed.) (1977b) *The Truly Cured Child: The New Challenge in Pediatric Cancer Care* (Baltimore, MD: University Park Press).

van Eys J. (1981) 'The Truly Cured Child: The Realistic and Necessary Goal in Pediatric Oncology', in Spinetta J.J. and Deasy-Spinetta P. (eds) *Living with Childhood Cancer*, pp. 30–40 (London: The C. V. Mosby Company).

van Eys J. (1985) 'Caring Toward Cure', *Child Health Care* 13:160–6.

van Eys J. (1987a) 'Ethical and Medicolegal Issues in Pediatric Oncology', *Hematology/Oncology Clinics of North America* 1:841–8.

van Eys J. (1987b) 'Living Beyond Cure: Transcending Survival', *The American Journal of Pediatric Hematology/Oncology* 9:114–18.

Waddington K. (2003) *Medical Education at St Bartholomew's Hospital 1123–1995* (Woodbridge: Boydell Press).

Wailoo K. (1997) *Drawing Blood: Technology and Disease Identity in Twentieth-Century America* (Baltimore, MD, and London: Johns Hopkins University Press).

Wailoo K. (2001) *Dying in the City of the Blues: Sickle Cell Anaemia and the Politics of Ethnicity and Health* (Chapel Hill, NC: University of North Carolina Press).

Wainwright M. (1990) *Miracle Cure: The Story of Penicillin and the Golden Age of Antibiotics* (Oxford: Blackwell).

Wakefield C.E., Mcloone J., Goodenough B., Lenthen K., Cairns D.R. and Cohn R.J. (2010) 'The Psychological Impact of Completing Childhood Cancer Treatment: A Systematic Review of the Literature', *Journal of Pediatric Psychology* 35:262–74.

Wall R. (2011) 'Using Bacteriology in Elite Hospital Practice: London and Cambridge, 1880–1920', *The Social History of Medicine* 24:739–57.

Ward G. (1917) 'Infective Theory of Acute Leukaemia' *British Journal of Children's Diseases* 14:10–20.

Watts S. (1997) *Epidemics and History: Disease, Power and Imperialism* (New Haven, CT, and London: Yale University Press).

Webster C. (2002) *The National Health Service: A Political History*, new ed. (Oxford: Oxford University Press).

Weindling P. (1992) 'From Infections to Chronic Disease: Changing Patterns of Sickness in the Nineteenth and Twentieth Centuries', in Wear A. (ed.) *Medicine in Society: Historical Essays*, pp. 303–16 (Cambridge: Cambridge University Press).

Weisz G. (2006) *Divide and Conquer: A Comparative History of Medical Specialization* (Oxford: Oxford University Press).

Whorton J. (2002) *Nature Cures: The History of Alternative Medicine in America* (Oxford: Oxford University Press).

Wilkinson S. and Kitzinger C. (2000) 'Thinking Differently About Thinking Positive: A Discursive Approach to Cancer Patients' Talk', *Social Science and Medicine* 50:797–811.

Williams I.G. (1942) 'Very High Voltage X-ray Therapy (Supervoltage)', *British Journal of Radiology* 15:360–4.

Williams I.G. (1946) 'Cancer in Childhood', *British Journal of Radiology* 19:182–97.

Williams I.G. (1957) 'The Role of Radiotherapy and Surgery in the Treatment of Cancer of the Breast: Presidential Address', *British Journal of Radiology* 30:505–15.

Wilson W.M., Farquhar J.W. and Lewis I.C. (1949) 'Procaine Penicillin in Infants and Children' *The Lancet* 1(6560):866–8.

Wiltshaw E. (2003) 'Leukaemia', in Christie D.A. and Tansey E.M. (eds) (2003) *Wellcome Witness To Twentieth Century Medicine Volume 15: Leukaemia*, 15–84 (London: The Wellcome Trust).

Woodgate R.L. (2006) 'Life is Never the Same: Childhood Cancer Narratives', *European Journal of Cancer Care* 15:8–18.

Worboys M. (1988a) 'Manson, Ross and Colonial Medical Policy: Tropical Medicine in London and Liverpool, 1899–1914' in MacLeod R. and Lewis M. (eds) *Disease, Medicine, and Empire: Perspectives on Western Medicine and the Experience of European Expansion*, pp. 21–37 (London: Routledge).

Worboys, M. (1988b) 'The Discovery of Colonial Malnutrition Between the Wars', in Arnold D. (ed.) *Imperial Medicine and Indigenous Societies*, pp. 208–25 (Manchester: Manchester University Press).

Worboys, M. (2004) 'Colonial and Imperial Medicine', in Brunton D. (ed.) *Medicine Transformed: Health, Disease and Society in Europe, 1800–1930*, pp. 211–38.

Wynne B. (1982) *Rationality and Ritual: The Windscale Inquiry and Nuclear Decisions in Britain* (Oxford: The British Society for the History of Science).

Zollman C. and Vickers A. (1999) 'ABC of Complementary Medicine. Users and Practitioners of Complementary Medicine', *BMJ* 319(7213):836–8.

Zubrod C.G. (1979) 'Historic Milestones in Curative Chemotherapy', *Seminars in Oncology* 6: 490–505.

Index

Printed in the United States
By Bookmasters